Advanced Polymer Concretes and Compounds

Oleg Figovsky
Dmitry Beilin

CRC Press
Taylor & Francis Group
Boca Raton London New York

CRC Press is an imprint of the
Taylor & Francis Group, an **informa** business

CRC Press
Taylor & Francis Group
6000 Broken Sound Parkway NW, Suite 300
Boca Raton, FL 33487-2742

Printed on acid-free paper
Version Date: 20131029

International Standard Book Number-13: 978-1-4665-9032-8 (Hardback)

Visit the Taylor & Francis Web site at
http://www.taylorandfrancis.com

and the CRC Press Web site at
http://www.crcpress.com

Contents

Foreword

At the beginning of the 21st century, nanoscale engineering became a driving force for discovery and innovation, not only in electronics and medicine, but also in materials and civil engineering. In the past 25 years, technology for corrosion protection has been advanced by the development of more corrosion-resistant structure materials and protective coatings. The authors of this book are well-known scientists in this area. Professor Oleg Figovsky is a leading inventor with more than 500 patents in the area of materials and civil engineering. The authors have published many articles on these topics in the last 10 years, but a complete book has not been published until now.

This book was written for engineers, students, and others who are interested in advanced materials. It reports the current status of advanced polymer and silicate polymer concretes and compounds. The scope of this book includes rubber concrete based on nanostructured polybutadiene binder, and silicate polymer concretes based on nanostructured organosilicate binder. It examines their physical, mechanical, and technological properties; their behavior upon exposure to harsh environmental factors; and the issues of durability and reliability. Additionally, the scope of this book includes novel polymer and silicate polymer coatings for corrosion and fire protection. One of the more important parts of this book is the chapter in which the authors present data regarding non-isocyanate polyurethane material for monolithic flooring and protective coating—the first nanostructured environment-friendly polyurethane coatings.

The emphasis in this book is on the service abilities of novel concretes and protective compounds for various environments, such as those involving water, pollutants, acid, and alkali substances. The book draws on much of the excellent research already performed on the durability and corrosion resistance of these materials.

Professor Vladimir Kestelman
kvnint@verizon.net

Preface

Developments in civil engineering and the growth of industry have created a continual demand for building materials with new and improved performance attributes. One of the current intensively progressing ways of improving efficiency of building structures is the use of a new class of building materials—polymer composites. It is now virtually impossible to find technical, transport, or building structures in which there are no composite materials. This progress is the result of the unique quality of these materials—a combination of high strength, at the level of structural steel—and the inherent features of nonmetallic materials.

New technological processes closely related with aggressive environments require increases in the manufacture of durable and effective composite materials that can withstand hostile media. Corrosion, the negative effects of radiation and temperatures, high UV radiation, and other adverse natural and anthropogenic effects on building structures are real problems that affect the human living environment.

A radical way to increase the durability of composite materials and products is the use of composites based on polymer binders. The first scientific results in this field were developed by Professor O. Figovsky and are protected by more than 25 patents in the United States, Germany, and Russia.

This book contains the descriptions and results of theoretical and experimental research in the field of efficient building material composites based on advanced polymer binders that were carried out by scientific teams from Polymate Ltd., International Nanotechnology Center (http://www.polymateltd.com, Israel) and Voronezh State University of Architecture and Civil Engineering (VGASU, Russia) with the direct participation or under the leadership of the authors. Physical and mechanical characteristics of these composites, including chemical resistance in various aggressive environments, are discussed in this book.

It is well known that polymer concrete (PC) is used in severe conditions in industrial and public buildings, as well as in transportation and hydraulic structures. The main advantages of polymer concrete over ordinary concrete are improved mechanical strength, low permeability, and improved chemical resistance. The main limitation is their relatively high material cost. For this reason, it is important to find the optimum technical–economic compromise.

This book examines the design issues related to the composition and properties of two new polymer concretes in relation to the polymer matrix and its material and building structure: rubber concrete based on polybutadiene binder and silicate polymer concrete with an organic–silicate matrix. Application of these polymer concretes in construction allows the builder to solve the problems of corrosion, the negative influence of temperature, degradation of a material at increased UV exposure, γ-radiation, and to increase the period between repairs, reliability, and durability of buildings and structures, especially those in aggressive environments. The complexities of physical–mechanical, heat–physical, and technological properties of these PCs, their behavior in environmentally aggressive conditions, and problems of durability and reliability are studied.

Special chapters are devoted to new environmentally friendly polymer compounds for monolithic industrial floor coverings and coatings. Well-known features of conventional polyurethane coverings are porosity, poor hydrolytic stability, and increased permeability. The involvement of toxic components, such as isocyanates, in the fabrication process make it extremely toxic and dangerous. New and promising methods for producing an epoxy–urethane hybrid compound allowed the material to be obtained with lower permeability, improved physical–mechanical characteristics, and safe manufacturing. The environmentally friendly two-component polyurethane binders for monolithic flooring and industrial coatings do not consist of isocyanate components at any stage of preparation, are insensitive to the moisture in the air or the coated surface, and have a number of advantages over conventional polyurethane materials.

Novel hydroxyurethane modifiers (HUM) for cold-cured epoxy composite materials were synthesized. It is established that the compositions with HUM demonstrate a significant increase in the speed of the curing process, a nontrivial increase in abrasion resistance, and a marked improvement in strength properties. The HUM, which possesses a wide range of hydrogen bonds, is embedded in an epoxy polymer network without a direct chemical interaction.

Advanced crack-resistant coatings based on water dispersion of chlorine-sulphopolyethylene (CSPE, Hypalon®) vulcanized by a Mannich alkali (MA) water solution were obtained. Application of MA as a CSPE structure component makes it possible to produce a vulcanized net of saturated polymer, and thus to develop an ecologically safe, impenetrable crack-resistant coating for any substrata (concrete, metal, plastic, etc.). The coatings can be applied in the aircraft, automotive, shipbuilding, paint, and varnish industries, civil engineering, and so on as a corrosion-resistant material. The optimal coating composition and its mechanical properties have been studied.

A new type of epoxy composition with nano-heterogenic structure based on epoxy resin, liquid rubber, amine hardener, and fluorinated surfactants of various chemical structures were developed. Formation of nano-heterogenic systems with fluorine-containing surface-active additives of optimal chemical composition is an effective method of obtaining advanced coatings. It has been shown that the mechanical properties and chemical resistance of nano-coatings are significantly higher with the use of surfactants, with the most effective surfactants being those with linear molecules containing carboxyl groups.

Acid-resistant building materials based on liquid glass find wide application in construction as silicate polymer concretes, filler pastes, putties, and so on. A significant increase in strength, heat, and fire resistance of the silicate matrix was achieved by introducing tetrafurfuryl esters of orthosilicic acid (tetrafurfuryloxisilane, or TFS) in the composition. Introduction of the TFS additive in the binding medium leads to the formation of the cross-linked polymer. The resulting nanostructured binder provided the basis for obtaining the acid-resistant silicate polymer concrete and void fillers.

The last chapter is devoted to the development of an advanced waterborne environmentally friendly and weather-resistant fire-protective coating composition. The composition consists of a combination of intumenscent organic and inorganic particles, an inorganic water glass, a water dispersion of chlorine-sulphonated polyethylene, and pigments and/or silicon dioxide. These fire-protective coatings are intended for indoor and outdoor application to flammable substrates such as wood, plastic, and so on. Standard laboratory tests of this coating composition have confirmed its excellent fire-protective of properties, corresponding to Class A fireproofing. The series of fire and heat-retardant coatings can have applications in the construction of wooden items and structures, plastic pipes, facings, and so on.

The major results of the works presented in this monograph were primarily published in the journal *Scientific Israel Technological Advantages* (http://www.sita-journal.com) from 2000 to 2013.

About the Authors

Professor Oleg Figovsky is the founder of the Polymate Ltd.-International Nanotechnology Research Centre (www.polymateltd.com), and its director of R&D, where he is working on research in nanostructured corrosion-resistant composite materials and protective coatings based on a polymer and silicate matrix. Novel nanotechnologies invented by Prof. O. Figovsky were the basis for establishing industrial production in the US, Canada, China, Mexico, Russia, and Israel. He is a member of the European Academy of Sciences, two Russian academies of sciences (RAASN and REA) and head of the UNESCO Chair "Green Chemistry." For his inventions in nanotechnologies he was awarded gold and silver medals at IENA-98 and the Gold Angel Prize at the Genius 2006 exhibition. Prof. O. Figovsky has authored books, more than 300 scientific articles and 500 patents.

Dr. Dmitry Beilin is head of the laboratory at the Polymate Ltd.-International Nanotechnology Research Center in Israel (http://www.polymateltd.com). His fields of interest include structural theory, thin-walled space structures, NDT diagnostics in industry and construction, and the strength of polymer composite materials. Dr. Beilin has authored more than 100 scientific articles and holds 5 patents.

Acknowledgments

The authors are happy to express profound gratitude to the research teams of Polymate Ltd.-INRC, including O. Aksenov, N. Blank, O. Birukova, V. Karchevsky, A. Leykin, R. Potashnikova, L. Shapovalov, and VGASU, led by Professor Yu. Borisov and Professor Yu. Potapov.

Special thanks to the CEO of Polymate Ltd-INRC, A. Trossman (Israel), and the CEO of Nanotech Industries, Inc., J. Kristul (http://www.nanotechindustriesinc.com, USA) for their assistance in the industrial application of scientific research results.

1 State of the Art in Polymer Concrete

NATURE OF POLYMER CONCRETE

Developments in civil engineering and industrial growth have created a continual demand for building materials with new and improved performance attributes. Nowadays, requirements to be met by construction materials include not only strength features but also chemical resistance, resulting from the increasing contamination of the natural environment, leading to the need to protect and increase the durability of building structures. Polymer concrete is an innovative and modern material that satisfies all the strict requirements of durability and chemical resistance, while offering high mechanical strength [1].

Polymer concrete (PC) is a composite material in which the binder consists entirely of a synthetic organic polymer. It is also known as *synthetic resin concrete*, *plastic resin concrete*, or simply *resin concrete*. Because the use of a polymer instead of Portland cement represents a substantial increase in cost, polymers should be used only in applications in which the higher cost can be justified by superior properties, low labor cost, or low energy requirements during processing and handling. It is therefore important that architects and engineers have some knowledge of the capabilities and limitations of PC materials in order to select the most appropriate and economic product for a specific application [2].

There are many situations for which polymer concrete proves to be the most appropriate material for the intended application, since conditions often dictate specific material requirements that may be met by PC when several composite properties are considered simultaneously.

The fast curing, excellent strength and durability, excellent damping properties, and wide range of elastic moduli available have made PC a very versatile material with many applications. Its primary disadvantages are high cost (binder cost ranges from less than $1 per pound to many dollars per pound); sensitivity of properties to temperature; volatility and flammability of monomers and resins; and lack of experience with PC by many users [3,4].

Polymer concrete consists of a mineral filler, for example an aggregate, and a polymer binder, which may be a thermoplastic, but more frequently is a thermosetting polymer (Figure 1.1).

When sand is used as a filler, the composite is referred to as a *polymer mortar*. Other fillers include crushed stone, gravel, limestone, chalk, condensed silica fume (silica flour, silica dust), granite, quartz, clay, expanded glass, and metallic fillers. Generally, any dry, nonabsorbent, solid material can be used as filler [4].

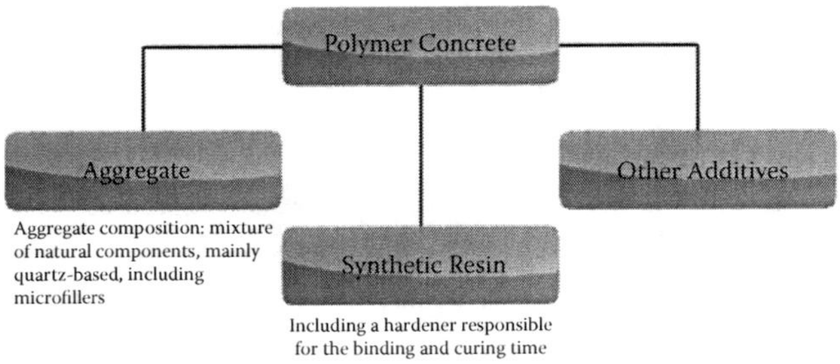

FIGURE 1.1 Composition of a polymer concrete. "Intended Use of Polymer Concrete," *Systemy i Technologie*, http://www.sytec.pl/en/polimerobeton-en.php#wlasciwosci-betonow-zywicznych.

Understanding of the nature of PC is necessary for the design of the most cost-effective PC composites and to produce materials with desired properties.

The polymer industry has made tremendous strides in the past 50 years, but it is quite likely that new and improved polymers will be developed [3], such as

- Polymers with more stable properties over a wide range of temperatures
- A much wider range of polymers that are compatible with fresh concrete
- Monomers, perhaps in the form of vapors, that can be used for producing PC much more rapidly and simply
- Resins that are designed to be recycled

COMPOSITION OF POLYMER CONCRETES

Polymer concrete (PC) is a composite material in which aggregates are bonded together with resins in a polymer matrix. Performance of PC is strongly dependent on various types and the mixed proportions of aggregates and resins [5].

Polymer concrete, as highly filled polymer compositions, can be prepared on any synthetic binding. However, due to the requirements for density, strength, deformability, chemical resistance, and other characteristics, about 10 different types of monomers or oligomers are used in practice. In combination with modifying additives, they provide more than 30 varieties of polymer concrete.

The polymer concretes are distinguished by the nature of the binder; e.g., furan, polyester, epoxy, phenol formaldehyde, carbamide, and so on. The classification of the main types of polymer concrete according to the kind of synthetic resins involved is shown in Figure 1.2 [7].

A variety of aggregate types have been used in PC silicates such as gravel, limestone, calcareous rock, granite, clay, quartz, crushed stone, silica sand or calcium carbonate ($CaCO_3$), as well as fine fly ash, phosphor-gypsum, cinder, and silica fume. Several silica sands have been used in the foundry industry [8]. Aggregates used must be usually dry and free of dirt to get the best bond between aggregates and resin. Figure 1.3 shows some aggregate systems used in PC mixtures [9].

The mix design of PC typically uses an aggregate size gradation to provide the lowest possible void volume and require the least polymeric binder necessary to coat

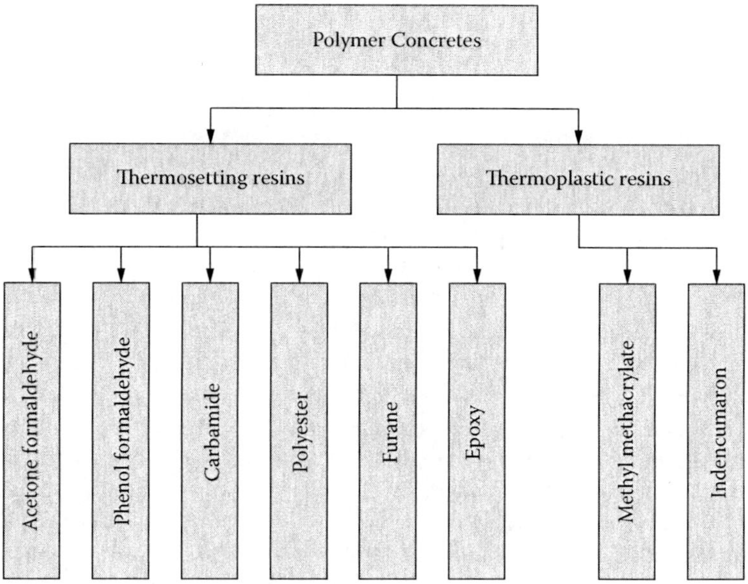

FIGURE 1.2 Classification of polymer concretes. (From V. Chmyhov, "Resistance of Rubber Concrete to Action of Aggressive Environments," doctoral thesis, Voronezh, 2002 [in Russian].)

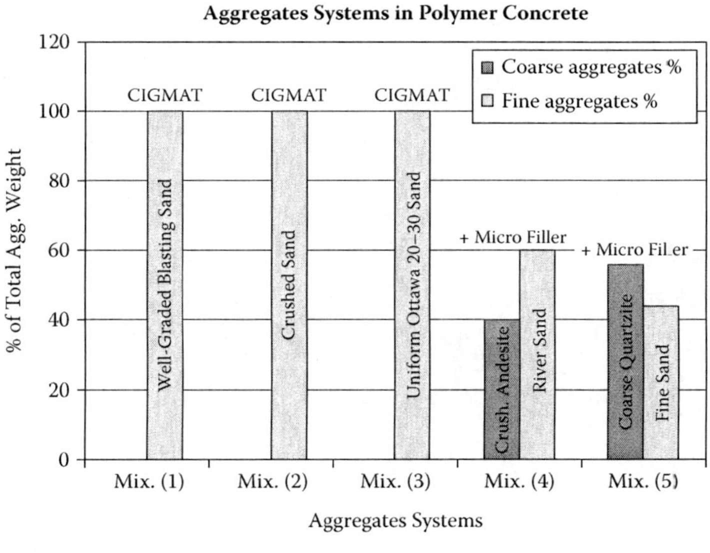

FIGURE 1.3 Most common aggregate systems used in PC. (From V. Y. Garas and C. Vipulanandan, "Review of Polyester Polymer Concrete Properties," http://www2.egr.uh.edu/~civeb1/CIGMAT/03_poster/11.pdf.)

the aggregates. One such combination is PC with 15 wt% (weight percentage) of resin and 85 wt% of siliceous sand with grain sizes of 245 mm to 342 mm.

To produce PC, a monomer or a prepolymer (i.e., a product resulting from the partial polymerization of a monomer), a hardener (cross-linking agent), and a catalyst are mixed with the filler. Other ingredients added to the mix include plasticizers and fire retardants. Sometimes silane coupling agents are used to increase the bond strength between the polymer matrix and the filler. Setting times for development of maximum strength can be readily varied from a few minutes to several hours by adjusting the temperature and the catalyst system. The amount of polymer binder used is generally small and is usually determined by the size of the filler. Normally the polymer content will range from 5% to 15% of the total weight, but if the filler is fine, up to 30% may be required [2,6].

To achieve the full potential of polymer concrete products for certain applications, various fiber reinforcements are used. Natural or synthetic fibers such as carbon or glass fibers can be added to the PC matrix to improve the mechanical performance of the polymer concretes. Mechanical improvement depends on the fiber type and on its concentration in the PC. For example, glass and organic fibers have little effect on the precracking behavior, but do substantially enhance the postcracking response, which leads not only to improved toughness and ductility, but also to higher tensile, flexural, and impact strength. The following are examples of the most commonly used fiber reinforcements [8,9]:

- Polyester fibers: improvement of 95% in the flexural strength have been obtained when adding 2% glass fibers and using silane as a coupling agent.
- Polyacrylonitrile (PAN) carbon fibers up to 6% (w/w) were used to reinforce PC. In compression, it increased the failure strain, but the strength and modulus decreased. In tension, the addition of carbon fibers increased failure strain, strength, and modulus [9].
- Glass fibers are noncorrosive, nonconductive, and nonmagnetic, and offer low density and high modulus. The addition of glass fibers enhanced flexural strength and toughness.
- Fracture properties: treatment with glass fibers further enhanced the flexural properties of PC.
- Different steel fibers were used; the effect of steel fibers in a PC system manifests as an increase in compressive, flexural, and impact strengths.

A special case of fiber reinforcement is the addition of oriented fibers. Reinforcement glass fibers and plastic bars placed along the principal stress directions reduce the creep deformation, which if present to a large degree, might result in an impaired structure or even cause structural collapse.

TYPES OF POLYMER CONCRETES

A wide variety of monomers and prepolymers are used to produce PC. The polymers most frequently used are based on five types of monomers or prepolymer systems:

1. Epoxy
2. Carbamide (urea-formaldehyde)

3. Acryl (methyl methacrylate)
4. Furan (furfuryl alcohol)
5. Polyester

The typical range of properties of PC products made with each of these polymers will be scrutinized in next part of this chapter.

Let us look more specifically at PC based on the kinds of binders previously listed.

Epoxy Polymer Concrete

Epoxy binder is a thermosetting polymer. The epoxy polymer can be hardened with a variety of curing agents, the most frequently used being polyamines (e.g., tertiary polyamines). The use of polyamine hardeners (curing agents) results in PC products with the highest chemical resistance. Other curing agents are polyamides and polysulfide polymers. Epoxy PC products cured with polyamides have greater flexibility, better heat resistance, and reduced chalking tendency in outdoor exposure, but their solvent and chemical resistance is lower than for similar products cured with polyamines. The adding of polysulfide polymers produces an epoxy PC with greater flexibility.

Epoxy PC exhibits high adhesion strength to most materials, low setting and postsetting shrinkage, high acid- and alkali-resistance (but nondurable in oxidizing medium), and good fatigue and creep resistance.

Because they are relatively expensive, epoxy polymers have not been used very widely as binders in PC products. Therefore, epoxy PC is used for special applications, in situations in which the higher cost can easily be justified, such as mortar for industrial flooring to provide physical and chemical resistance, skid-resistant overlays (filled with sand, emery, pumice, quartz) in highways, epoxy plaster for exterior walls (e.g., in exposed aggregate panels), and resurfacing material for deteriorated areas (e.g., in flooring). Epoxy PC reinforced with glass, carbon, or boron fibers is used in the fabrication of translucent panels, boat hulls, and automobile bodies [2,6].

Dilution of epoxy resins is achieved by introducing solvents. In this case, the viscosity is much reduced and the pot life is increased. However, the use of volatile solvents leads to formation of pores and capillaries in the binder, and hence to a reduction in polymer density [7].

It was concluded that the amount of resin and filler in the chemical composition of the fabricated epoxy PCs had a great influence on identification of the maximum physical strength. The PC specimen with 15% resin and 200% filler resulted in maximum compressive and flexural strength. The tensile strength was maximized with 20% resin and 200% filler. The mechanical strength of fabricated PCs was 4- to 5-fold higher than Portland cement concrete [5].

Carbamide Polymer Concrete

Carbamide (urea-formaldehyde) resin is prepared by the poly condensation reaction of urea and formaldehyde in aqueous or aqueous-alcoholic medium.

Polymer concretes based on carbamide resins have a low toxicity and are favorable in manufacturing. However, the content of the polymer matrix in these PCs is high (up to 30%) and their physical and mechanical properties are low in comparison to other types of polymer concretes.

It should be noted that carbamide resins contain a large amount of tree water (30%–40%), which leads to more shrinkage during curing of the composition. In some cases this leads to cracking of the material [7].

Acryl Polymer Concrete

The most common acrylic polymer is polymethyl methacrylate, which is obtained by polymerization of methyl methacrylate (MMA). The material consists of a highway-grade aggregate and a matrix produced by cross-linking MMA with trimethylol propane trimethacrylate or other polyfunctional acrylic oligomers.

PC made with this acrylic polymer as a binder is a versatile material, has excellent weathering resistance, good waterproofing properties, good chemical resistance, and relatively low setting shrinkage (0.01% to 0.1%); its coefficient of thermal expansion is equivalent to that of Portland cement concrete. Because of a very low tendency to absorb water, acrylic PC has a very high freeze–thaw resistance. The low flash point (11°C) of the MMA monomer is a disadvantage, however, as it constitutes a safety problem.

Although the MMA monomer is more expensive than the prepolymer monomer used in the more popular polyester PC, its unique properties account for its use in a great many diverse applications, including the manufacture of stair units, sanitary products, curbstones, and facades. The low temperature capability and long-term durability make the material ideal for emergency repairs that must be accomplished when standard concrete repair materials cannot be used. A highly successful development has been its use as a rapidly curing, structural patching material for repairing large holes in bridge decks, spall repairs, emergency full-depth repairs, bridge expansion joint headers, bearing pads, closure pours, and concrete structural beam repairs. The MMA polymer concrete can be applied at any time after the primer has cured, which allows for preparation of many areas ahead of time making placement more efficient [2,6,10].

Furan Polymer Concrete

Furan polymers are based on furfuryl alcohol, which is derived from agricultural residues such as corncobs, rice hulls, oat hulls, or sugar cane bagasse. The furan prepolymer is usually cross-linked with furfuryl alcohol, furfuraldehyde, or formaldehyde to yield thermosetting polymers, highly resistant to most aqueous acidic or basic solutions and strong solvents such as ketones, aromatics, and chlorinated compounds. The important characteristic of furan resins is their ability to be stored for long periods of time (up to 5 years), even at low temperatures.

However, short-term pot life and toxicity in the uncured state hamper the use of these resins, and high self-heating temperature causes significant thermal stresses, which adversely affect the strength characteristics of furan PC. It was found that the degradation of furan polymers due to aging leads to a decrease in the mechanical properties of the composite.

The furan polymers are used as binders in mortars and grouts to achieve chemically resistant brick floors (e.g., carbon brick and red shale brick) and linings. In addition to exhibiting superior chemical resistance, these floors have excellent resistance to elevated temperatures and extreme thermal shock [2,6,7].

Polyester Polymer Concrete

Polyester resins like epoxy are a type of thermosetting resins, obtained by polycondensation. They have a low viscosity, and materials that are based on them have high mechanical and electrical insulating properties and high resistance to acids, gasoline, oils.

Because of low cost, the most widely used unsaturated polyester polymer is in the form of 60% to 80% solutions of the prepolymer in copolymerizable monomers such as styrene and a mix of styrene with methyl methacrylate. During hardening, the polyester prepolymer and the monomer react through their unsaturated groups (double bonds).

It should be added that polyester resins are toxic, and with poor stirring of the mixture components the probability of stratification and level-by-level curing of the composition is high.

Polyester PC has high mechanical, cohesion and adhesion strength, and good chemical and freeze–thaw resistance. It has, however, large setting and postsetting shrinkage (up to 10 times greater than Portland cement concrete), which is a serious disadvantage in certain applications.

Polyester PC is used in various procast and cast-in-place applications in construction works, public and commercial buildings, for the manufacture of sanitary engineering, cladding of composite pipes in bathrooms and toilets, stairs, and chemical-resistant floors [2,6,7].

Other Types of Polymer Concrete

Polymer concretes based on phenol-formaldehyde, acetone-formaldehyde resins and monomers, and methyl methacrylate are much less common. Phenolic resins are similar to furan in many physical and mechanical properties. However, they are unstable in alkalis like polyester resins [7].

PHYSICAL–MECHANICAL PROPERTIES OF POLYMER CONCRETE

Using synthetic resins instead of traditional Portland cement binder allows concrete to be created with a series of interesting properties such as high chemical resistance to many corrosive environments or high mechanical strength. In ordinary concrete, the strength properties of cured cement paste are at least several times lower than the corresponding features of the mother rocks of the aggregate, and the adhesion of binder and aggregate is relatively low. The situation is different in resin concretes: the tensile strength of hardened resin binder is much higher, and the compressive strength is similar to the strength of the rocks from which the aggregate was obtained.

Polymer concrete composites have generally good resistance to attack by chemicals and other corrosive mediums, have very low water sorption properties, good resistance to abrasion, and marked freeze–thaw stability. Also, the greater strength of polymer concrete in comparison to that of Portland cement concrete permits the use of up to 50% less material. This puts polymer concrete on a competitive basis with cement concrete in certain special applications. The chemical resistance and physical properties are generally determined by the nature of the polymer binder to a greater extent than by the type and the amount of filler. In turn, the properties of the matrix polymer are highly dependent on time and the temperature in which it is exposed.

The viscoelastic properties of the polymer binder give rise to high creep values. This is a factor in the restricted use of PC in structural applications. Its deformation response is highly variable depending on the formulation, and the elastic module may range from 20 to about 50 GPa, tensile failure strain being usually 1%. Shrinkage strains vary with the polymer used (high for polyester and low for epoxy-based binder) and must be taken into account in an application [1,2,6].

Other advantages of polymer concrete should be mentioned as well [1]:

- Impervious to liquids, small number of pores, absolute lightness
- High freeze–thaw resistance resulting from non-moisture-absorbing property; good electric insulation
- High resistance to corrosive chemical substances, including acids and bases
- High resistance to scratches; it does not peel, does not splinter, does not require any maintenance, and experiences no erosion, which reduces costs of maintenance and exploitation
- Increased flexural, compressive, and tensile strengths; fast setting times (curing within 1 or 2 hours); good durability
- May be applied for any load class
- Due to the properties of polymer concrete, products made of this material are durable and strong and are characterized by higher mechanical resistance to loads than traditional concrete, which means that the cross-section area for comparable load classes is smaller in polymer concrete products—thus they are lighter than concrete products, which results in easier and quicker installation.
- Due to natural components, it is an environment-friendly material, debris may be reused (aggregate can be returned to the production process)
- Good adhesion to essential construction materials (steel, traditional concrete)
- Good ability to dampen vibrations due to resins contained in the material
- Possibility to obtain very smooth surfaces, guaranteeing many practical applications
- Possibility to design eye-catching, durable color solutions (according to the RAL–colors chart* do not fade for a very long period of time—resistant to UV radiation)

* Reichs-Ausschuss für Lieferbedingungen (German).

- Resistance to changing weather conditions and atmospheric factors
- Very short time to achieve installation and usage efficiency
- Easy drilling and cutting with diamond bits and saw blades
- High abrasion resistance (comparable to granite)

The advantages of polymer concrete are particularly noticeable when comparing its individual properties to traditional B30 class concrete (Table 1.1).

One can see that epoxy polymer concrete exhibits higher compressive strength, with values ranging from 3.8–4.3 times higher than commercial concrete. Compressive strength of this PC shows an increase of 1.5 times when glass fibers are added to the epoxy resin matrix and 2.0 times with the use of carbon fibers [8].

Corrosion Resistance of Polymer Concretes

Nonreactivity or corrosion (chemical) resistance of polymer concretes to aggressive influences is one of the most important features of their physical and mechanical properties. It is defined as the ability of a material to resist corrosive environments and preserve the properties and forms.

The action of an aggressive environment on polymer concretes depends on their permeability and subsequent dissolution and elution of the binder components by hydrolysis of molecular bonds and the weakening of the interaction between the polymer and filler.

The experience of studying the corrosion resistance of some polymer concretes is summed up below.

Furan polymer concrete is a corrosion-resistant polymer material. Endurance tests (a year or more) showed high resistance of this material to most industrial chemicals, except oxidants (nitric acid, acetic acid) and some solvents (acetone, benzene, alcohol).

Furan polymer concrete consistently maintains its properties in solutions of sulfuric and hydrochloric acids, chlorides, alkalis, as well as in fats, sugars, oils, and petroleum products. It is noted that furan polymer concrete has high resistance to aggressive action of sulfuric acid at high concentrations and temperatures. The reason is that the furan resins are partially cured with acid and at the same time, water is a more aggressive environment than the acid. Therefore, the higher the concentration of the acid, the higher the resistance of this polymer is. Furan PCs have good resistance to the action of alkaline environments at low temperatures and poor resistance at temperatures above 50°C. Chemical resistance of furan PCs to alkalis depend on the kind of filler; coke filler provides a high resistance [7].

Polyester *polymer concrete* resists oxidizers, acids, oils, and petroleum products, but there is not enough resistance to alkaline solutions and water. The strength of polyester PC in water decreases faster than in solutions of inorganic salts and some acids; therefore, resistance in acid solutions can simultaneously serve as an estimate of water resistance. For example, the flexural strength of the PC, immersed in a 10% solution of sulfuric acid or 10% sodium chloride solution is reduced by 30% after 80 days of immersion.

TABLE 1.1
Physical–Mechanical Properties of Polymer Concretes

Indicator	Unit	Kind of Polymer Concrete						Portland Cement Concrete (B30)
		Epoxy	Furan	Furan-Epoxy	Polyester	Carbamide	Acrylate	
Average density	10^2 kg/m^3 kg/m^3	22–24	22–24	22–24	22–24	22–24	22–24	**19–25**
Compression strength	MPa	50–150	70–90	90–110	80–100	40–70	70–90	**13–35**
Tension strength		10–40	5–8	9–11	7–9	3–7	10–13	**1.5–3.5**
Flexural strength		15–50	—	—	15–45	—	30–33	**2–8**
Modulus of elasticity	10^4MPa	3–4	2–3.2	3.2–3.8	1.8–3.6	1–1.5	1–1.5	**2–3**
Poisson's ratio	—	0.24–0.29	0.32	0.27	0.2–0.23	0.22–0.26	0.27	**0.15–0.2**
Shrinkage at hardening	%	0.005–0.09	0.1	0.05–0.08	0.08–0.1	0.2–0.25	0.15–0.2	**0.015–0.3**
Water sorption		—	0.05–0.9	0.01	0.05–+0.1	0.2–0.3	0.01	**5–+8**
Heat resistance	°C	120–130	120–140	120	80	100–110	60	**200–400**
Wear resistance	kg/m^2	0.05–0.1	0.18–0.21	0.05–0.1	0.15–0.25	0.2–03	—	**0.85**
Freeze resistance	Number of cycles	500	300	500	300	200	500	**1000**
Toughness	J/cm^2	3–10	0.15–0.25	0.16–0.25	0.2–0.25	0.15–*0.25	—	**0.5**

Source: Data from V. Chmyhov, "Resistance of Rubber Concrete to Action of Aggressive Environments," doctoral thesis, Voronezh, 2002 (in Russian).

PCs based on epoxy resin binder are resistant to many corrosive environments such as a concentrated solution of caustic soda at 90°C; the action of alkali metal salts, alcohols, oils, gasoline; and other aliphatic hydrocarbons. The flexural strength of epoxy PC is slightly affected by immersion in a 10% solutions of sulfuric acid and chloride sodium, which is an indicator of the good chemical resistance of this kind of concrete to these aggressive agents [11]. However, these polymer concretes are not resistant to a sulfuric acid concentration of 60% and nitric acid concentrations above 70% [12].

Water Resistance of Polymer Concretes

Corrosive attack of a water environment on polymer concrete is manifested as a change in its structure and properties without disruption of integrity or with destruction of the materials. Water penetrates between macromolecules through micropores and fine capillaries and irreversibly changes the chemical structure of the polymer binder. Such changes result in a lowering of physical–mechanical properties of the material, its destruction, cracking, and so on. So physical and mechanical properties of polymer concrete products and structures depend heavily on the humidity of the operating environment.

Following are the results of long-term experimental research on the effect of a water environment on the physical and mechanical characteristics of polymer concretes [14].

Polymer concrete samples of four compositions (Table 1.2) measuring 4 × 4 × 16 cm are prepared. Control samples were placed in a desiccator containing moisture absorption silica gel composition. Other samples were in environments with various humidity. The influence of environmental humidity on the compressive strength, modulus elasticity, and moisture absorption of furfurol–acetone, epoxy, polyester, and rubber (RubCon) polymer concrete samples are shown in Tables 1.3–1.5 and Figures 1.4 and 1.5.

Furfurol–Acetone Polymer Concrete [12,14,16]

The ultimate compressive strength of furfurol–acetone polymer concrete control samples is 65 MPa. The most intensive decrease in the strength of the samples stored in environments with humidity of 50% to 60% and 86% to 96%, starts with 110 days (Figure 1.6).

After 350 days of exposure in the environment with humidity of 50% to 60%, compressive strength of the samples is reduced by 10%, and for the samples stored in 86% to 96% humidity, compressive strength is reduced by 18%. The greatest decrease in durability is noted at the maximal saturation of the polymer concrete samples by water. The first 20-day decrease in strength of the samples reaches 10% (i.e., approximately the same as at 50% to 60% humidity for 350 days of storage). In the next 80 to 90 days, the compressive strength of the samples decreases sharply. Strength reduction is then slowed and, after about six months, stops. Further exposure of samples in water does not change their strength, which is 64% of the initial value.

TABLE 1.2
Composition of the Polymer Concrete Samples [15]

Ingredients	Composition of the Polymer Concrete Samples (weight parts)			
	Furfurol–Acetone	**Epoxy**	**Polyester**	**Rubber (RubCon)***
Furane resin	100	—	—	—
Epoxy resin	—	100	—	—
Polyester resin	—	—	100	—
Liquid polybutadiene	—	—	—	100
Benzene sulfonic acid	25	—	—	—
Polyethylene polyamine	—	10	—	—
Isopropylbenzene hydroxide	—	—	4	—
Cobalt naphthenate	—	—	8	—
Sulfur	—	—	—	51
Tetramethylthiuramdisulfide (Thiuram)	—	—	—	4.5
Mercaptobenzothiazole (Captax)	—	—	—	0.6
Zink oxide	—	—	—	18.8
Calcium oxide	—	—	—	5
Microfiller	100	100	100	88
Sand	290	100	200	300
Crushed stone	540	600	600	682.2

Source: Reprinted from Yu. Borisov, Yu. Potapov, O. Figovsky, D. Beilin, "Water Resistance of the Polymer Concretes," *J. Scientific Israel Advanced Technology* 14, no. 3 (2012): 84–91. With permission.

For more detailed information about the composition and properties of RubCon, see Chapter 2.

TABLE 1.3
Compressive Strength of the Polymer Concretes (Time of exposition in water environment is one year.)

No.	Polymer Concrete	Compressive Strength, MPa		
		Control Samples	**After Exposition in Water**	**After Drying**
1	Furfurol–acetone	65	42	55
2	Epoxy	92	70	85
3	Polyester	86	47	51
4	Rubber (RubCon)	94.8	94.7	94.8

Source: Reprinted from Yu. Borisov, Yu. Potapov, O. Figovsky, D. Beilin, "Water Resistance of the Polymer Concretes," *J. Scientific Israel Advanced Technology* 14, no. 3 (2012): 84–91. With permission.

TABLE 1.4
Modulus of Elasticity of the Polymer Concretes (Time of exposition in water environment is 1 year.)

		Modulus of Elasticity, MPa		
No.	**Polymer Concrete**	**Control Samples**	**After Exposure in Water**	**After Drying**
1	Furfurol–acetone	19000	6300	8200
2	Epoxy	22000	13500	18600
3	Polyester	21000	6000	8000
4	Rubber (RubCon)	25800	23800	25000

Source: Reprinted from Yu. Borisov, Yu. Potapov, O. Figovsky, D. Beilin, "Water Resistance of the Polymer Concretes," *J. Scientific Israel Advanced Technology* 14, no. 3 (2012): 84–91. With permission.

Deformation properties of the furfurol–acetone polymer samples are reduced to a much greater extent. The modulus of elasticity of the polymer concrete samples starts to decrease noticeably (about 10% to 15%) at 50% to 95% humidity after just 150 days; for the samples in water, a 30% reduction of the modulus has been observed after 50 days (Figure 1.7). After 350 days, the modulus of elasticity is reduced by 28% for the samples in the environment with 50% to 60% humidity, 46% in the medium with 85% to 95% humidity, and 67% for the samples stored in water. As this takes place, the trend of modules is not identical: the modulus of elasticity of the first two series of samples trends downward over time, whereas for the samples in water, this tendency stopped after 6 months of exposure.

A noted phenomenon is caused by water absorption of polymer concrete (Figure 1.8). The penetration of water into polymer concrete leads to a softening of its structure and consequently to loss of strength and an almost threefold increase in the deformability of the material.

Water penetration of the test samples occurs in several ways. Samples that are placed in mediums with humidity of 50%–60% and 85%–95% swell gradually and are not stabilized in weight within one year. The samples stored in water are saturated with it in half a year and their subsequent storage in these conditions does not lead to increase in weight. By this time, the stress–strain parameters in compression are stabilized. Note that these parameters continue to decrease for the samples in 50%–60% and 85%–95% humidity because water absorption of polymer concrete continues.

It is apparent that a water environment is a limiting case. As this takes place, the ultimate compressive strength and the module of elasticity of polymer concrete samples decreased accordingly, up to 44,000 and 63,000 MPa.

Let's consider the reasons for decreases in the stress–strain parameters of furfurol–acetone polymer concrete in a wet environment. The process of water absorption by polymer concrete occurs in two ways: by diffusion and absorption. The latter is associated with the presence of free benzene sulfonic acid, which is highly hygroscopic and absorbs moisture from the environment and thus increases

TABLE 1.5
Characteristics of Polymer Concretes Depend on Time of Exposure in Water Environment

Time (days)	Water Absorption (%)			Compressive Strength (MPa)			Modulus of Elasticity (MPa)		
	Epoxy	Polyester	Rubber (RubCon)	Epoxy	Polyester	Rubber (RubCon)	Epoxy	Polyester	Rubber (RubCon)
0	0	0	0	92	86	94.9	22000	21000	25800
1	0.001	0.001	0.001	92	85	94.8	21500	20500	24500
7	0.005	0.01	0.002	91	82	94.8	21000	19000	24300
30	0.05	0.01	0.03	86	75	94.7	20000	16000	24150
90	0.1	0.5	0.44	80	60	94.7	18000	10000	24000
180	0.2	0.7	0.45	76	55	94.7	16000	8600	24000
1 year	0.25	0.91	0.45	72	49	94.7	15000	7000	24000
2 year	0.3	0.92		70	48		14000	6500	
3 year	0.31	0.93		70	48		13500	6000	

Source: Reprinted from Yu. Borisov, Yu. Potapov, O. Figovsky, D. Beilin, "Water Resistance of the Polymer Concretes," *J. Scientific Israel Advanced Technology* 14, no. 3 (2012): 84–91. With permission.

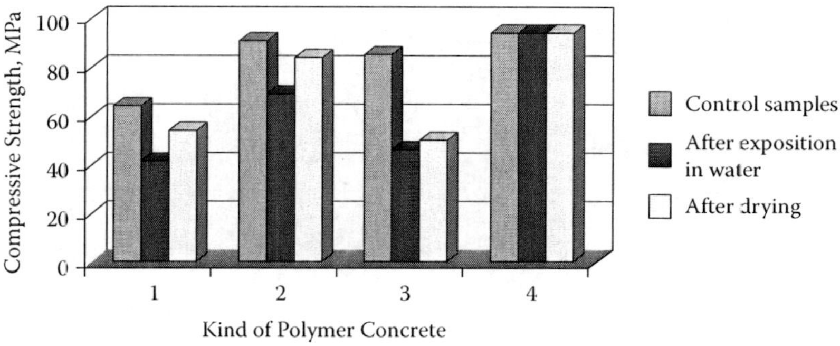

FIGURE 1.4 Compressive strength of the polymer concretes in a water environment. (From Yu. Borisov, Yu. Potapov, O. Figovsky, D. Beilin, "Water Resistance of the Polymer Concretes," *J. Scientific Israel Advanced Technology* 14, no. 3 (2012): 84–91. With permission.)

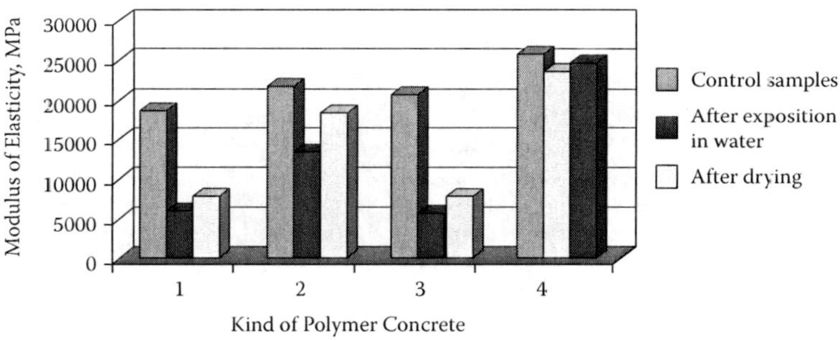

FIGURE 1.5 Modulus of elasticity of polymer concretes in a water environment. (From Yu. Borisov, Yu. Potapov, O. Figovsky, and D. Beilin, "Water Resistance of the Polymer Concretes," *J. Scientific Israel Advanced Technology* 14, no. 3 (2012): 84–91. With permission.)

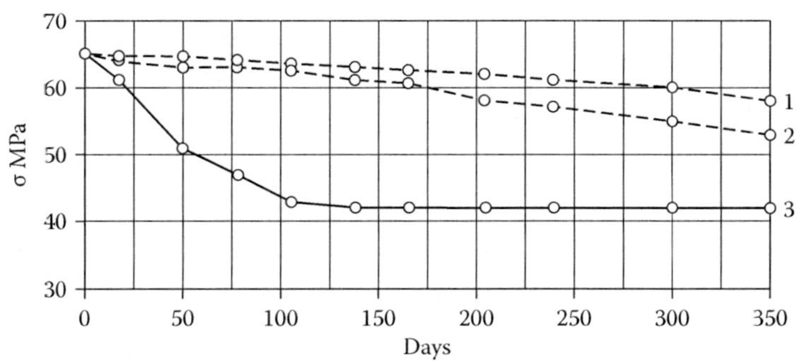

FIGURE 1.6 Dependence of ultimate compressive strength of furfurol–acetone polymer concrete samples on exposition time at humidity of environment: 1: 50%–60%, 2: 85%–95%, 3: water immersion. (From Yu. Borisov, Yu. Potapov, O. Figovsky, D. Beilin, "Water Resistance of the Polymer Concretes," *J. Scientific Israel Advanced Technology* 14, no. 3 (2012): 84–91. With permission.)

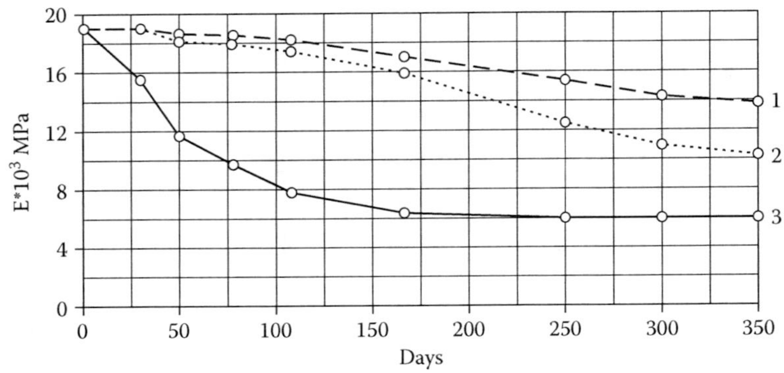

FIGURE 1.7 Dependence of modulus elasticity of furfurol–acetone polymer concrete samples on exposition time at humidity of environment: 1: 50%–60%, 2: 85%–95%, 3: water immersion. (From Yu. Borisov, Yu. Potapov, O. Figovsky, and D. Beilin, "Water Resistance of the Polymer Concretes," *J. Scientific Israel Advanced Technology* 14, no. 3 (2012): 84–91. With permission.)

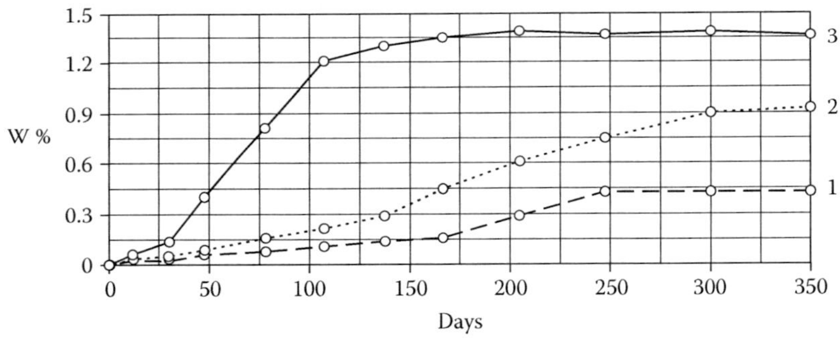

FIGURE 1.8 Dependence of water absorption of furfurol–acetone polymer concrete samples on exposition time at humidity of environment: 1: 50%–60%, 2: 85%–95%, 3: water immersion. (From Yu. Borisov, Yu. Potapov, O. Figovsky, and D. Beilin, "Water Resistance of the Polymer Concretes," *J. Scientific Israel Advanced Technology* 14, no. 3 (2012): 84–91. With permission.)

the water content in the polymer. Hydrophilicity of the mineral filler also promotes absorption of moisture. On completion of this process, the compressive strength and the module of elasticity decrease, but eventually are stabilized. Reducing the strength of the polymer concrete causes a weakening of the adhesive bond forces due to the formation a water layer between the resin and the filler surface.

Epoxy and Polyester Polymer Concrete [12,14,16]

Experiments similar to those previously described were carried out with epoxy and polyester polymer concrete samples (compositions 3 and 5, Table 1.2).

The ultimate compressive strength of drying epoxy polymer concrete samples is about 92% of the control sample strength, which testifies to a high degree of reversibility of the process associated with the adhesive nature of the interaction between binder and filler.

The maximal decrease in compressive strength (up to 40%) is observed in the polyester composites, which is caused by destruction of the bonds between polyester molecules and between binder and filler.

Rubber Polymer Concrete (RubCon) [7,17–20]

The samples of rubber polymer concrete (RubCon) (composition 5, Table 1.2), sized 4 × 4 × 16 cm, were tested under conditions similar to those previously described.

The ultimate compressive strength of the RubCon samples is 94.8 MPa. It is shown in Figure 1.9 that the strength of the concrete samples decreases moderately regardless of the environment's humidity. After a 50-day exposure, the strength limit stabilizes and remains unchanged until the end of the test. This can be explained by low water absorption of the rubber polymer concrete, not exceeding 0.05%.

Deformation properties of rubber concrete remain unchanged at humidity of 50% to 60%. By increasing the humidity up to 85% to 95%, a slight decrease in elastic modulus was observed at the initial stage of exposure and it stabilized after 50 to 60 days. Changes in stress–strain characteristics of samples in water were the same. An insignificant decrease and subsequent stabilization was noted in the elastic modulus (up to 3.5%) with an exposure of up to 50 days (Figure 1.10). The bending point corresponds to the total water absorption of the rubber concrete samples, which is terminated after 50 days of exposure (Figure 1.11).

Drying of RubCon samples does not have a significant influence on its strength and deformation parameters due to a very small quantity of the adsorbed moisture (water saturation is 0.045%).

The tests have shown that rubber polymer concrete (RubCon) is the composite with the highest water resistance. During the exposure of the RubCon samples in

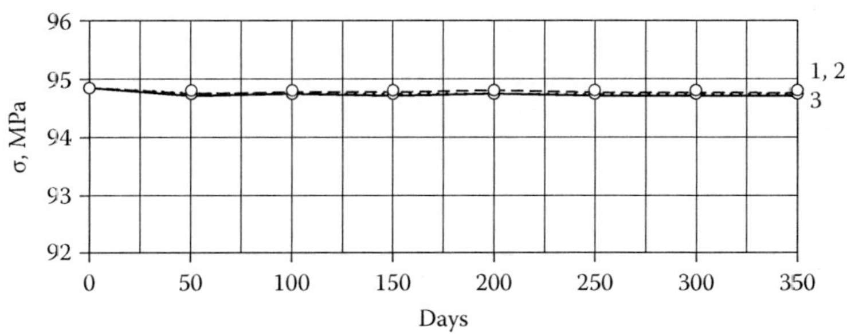

FIGURE 1.9 Dependence of compressive strength of rubber polymer concrete (RubCon) samples on exposition time at humidity of environment: 1: 50%–60%, 2: 85%–95%, 3: water immersion, (From Yu. Borisov, Yu. Potapov, O. Figovsky, and D. Beilin, "Water Resistance of the Polymer Concretes," *J. Scientific Israel Advanced Technology* 14, no. 3 (2012): 84–91. With permission.)

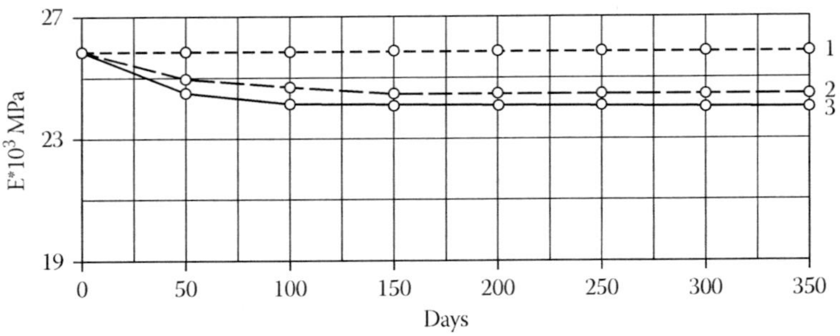

FIGURE 1.10 Dependence of modulus elasticity of rubber polymer concrete (RubCon) samples on exposition time at humidity of environment: 1: 50%–60%, 2: 85%–95%, 3: water immersion. (From Yu. Borisov, Yu. Potapov, O. Figovsky, and D. Beilin, "Water Resistance of the Polymer Concretes," *J. Scientific Israel Advanced Technology* 14, no. 3 (2012): 84–91. With permission.)

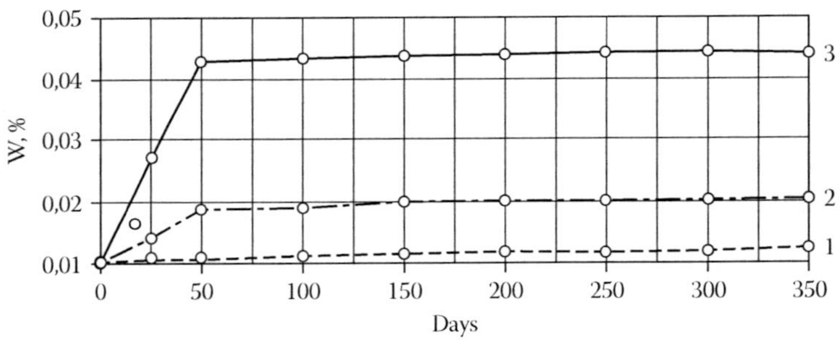

FIGURE 1.11 Dependence of water absorption of rubber polymer concrete (RubCon) samples on exposition time at humidity of environment: 1: 50%–60%, 2: 85%–95%, 3: water immersion. (From Yu. Borisov, Yu. Potapov, O. Figovsky, and D. Beilin, "Water Resistance of the Polymer Concretes," *J. Scientific Israel Advanced Technology* 14, no. 3 (2012): 84–91. With permission.)

a water environment, their strength remained almost unchanged; water absorption was only 0.045%. Such a small mass change can be attributed to the hydrophobicity of the material and nonpolarity of the rubber binder.

INTENDED USE OF POLYMER CONCRETES

Polymer concrete is usually used in severe conditions in industrial and public buildings as well as in transportation and hydraulic structures. The main uses are repair/strengthening, and corrosion protection of concrete structures. The main advantages of polymer concrete over ordinary concrete are improved mechanical strength, low permeability, and improved chemical resistance. A highly successful development has been its use as a rapid-setting, structural patching material for repairing large holes in bridge decks [6]. One of the advantages of polymer concrete is its ability to dampen vibrations. It had been used mainly in foundations that support high-frequency equipment, but now it is being used for building structures undergoing dynamic loads [13,14].

Precast PC has been widely used for flooring, utility boxes, manholes, drains, wall panels, and machine bases. Precast PC can be produced quickly, as compared to precast concrete or machined cast iron, has a high strength-to-weight ratio, and can be produced with cast inserts. However, the limited pot life of polymeric binder mix makes it difficult to obtain high-quality large products and structures made from epoxy, polyester, furan, and carbamide polymer concretes. Significant shrinkage of epoxy PC and large deformability of polyester PC lead, in some cases, to defects in the manufacturing and operation of products and structures made of these polymer concretes.

The main limitation of the wide application of PCs is their relatively high material cost. This is why it is important to find the optimum technical–economic compromise [8].

Due to its properties, polymer concrete is currently used in many applications:

- In highway pavements
- As underground wastewater pipes
- For manufacturing thin overlays
- As precast components for bridge panels, overlay bridge decks, bridge drainage systems (bridge edge beams, bridge curbs, bridge drainage inlets, gutters), for linear drainage systems (channels, linear drainage channels, linear drainage silt boxes), parking garage decks, industrial floors and dams, buildings, machine bases, and transportation components, as well as for hydraulic structures such as dams, dikes, reservoirs, and piers (where PC creates a highly abrasion-resistant surface)
- In high-pressure and high-temperature environments, such as the generation of geothermal energy, due to its durability in hot acidic springs
- For production of industrial tanks intended for electrolysis of nonferrous metals, production of catch basins and channels to drain aggressive industrial wastewater, water meter chambers, sewage pump stations, production of storage tanks to store corrosive substances such as acids and bases
- For repair of structures and for coatings, due to its strong bonding with PCC (there must be no sealants, curing compounds, or any other material covering the concrete, as it will interfere with the adhesion of PC) [3,4,8]

The most effective scopes of polymer concrete application, depending on the type of polymeric binder, are shown in Table 1.6.

In conclusion, it should be emphasized that the main advantage of polymer concretes is high chemical resistance characterized by long-term resistance to corrosive environments. Composition based on polymer binders are the most promising for use in these conditions.

Improving the efficiency of polymer composites is possible by applying cheaper binders and eliminating waste during chemical production by reducing the components used, decreasing the toxicity of polymeric composition, and increasing the degree of polymerization [7].

TABLE 1.6
Fields of Use of Some Polymer Concretes

Type of the Binder	General Characteristics	Typical Application
Polymethyl methacrylate	• Low water absorption • High freeze–thaw resistance • Low rate of shrinkage • Good chemical resistance • Outdoor durability	• Stair units • Façade plates • Sanitary products for curbstones
Polyester	• Good adhesive properties • Good chemical and freeze–thaw resistance • High-setting and postsetting shrinkage	• Panels for public and commercial buildings • Floor tiles • Pipes • Stairs • Precast and cast-in applications in construction works
Epoxy	• Good adhesive properties • Low shrinkage • Superior chemical resistance • Good creep and fatigue resistance • Low water sorption	• Mortar for industrial flooring • Skid-resistant overlays in highways • Epoxy plaster for exterior walls and resurfacing of deteriorated strictures
Furan based polymer	• High resistance to chemicals (most acidic or basic aqueous media)	• Floor and linings that are resistant to chemicals, elevated temperatures, and thermal shocks

Source: A. Blaga and J. J. Beaudoin, *Polymer Concrete*, CBD-242, 1985, http://web.mit.edu/parmstr/Public/NRCan/CanBldgDigests/cbd242_e.htm.

ADVANCED POLYMER CONCRETES BASED ON NOVEL BINDERS

Rubber Concrete with Vulcanized Polybutadiene Binder [17,20]

Rubber concrete (RubCon) is a new polymer composite material created and studied by the research team of the Voronezh State University of Architecture and Construction (Russia) and the Israel company Polymate Ltd.-1NRC.

RubCon is based on low molecular polybutadiene containing up to 90% mineral filler by weight. It does not contain cement as a binder, and therefore has elastic properties, the highest resistance to chemical corrosion, is highly repellant to water, and has remarkable compression strength (up to 100 MPa). RubCon does not exhibit the common failure mechanisms of conventional concrete such as cracking and flaking, and freeze and thaw. RubCon has high vibration resistance that makes it ideal pad material for chemical pumps and reactor foundations.

RubCon is produced using a special high-speed extruder mixer and is formed as a conventional polymer concrete with following hot dry air vulcanization (Figure 1.12).

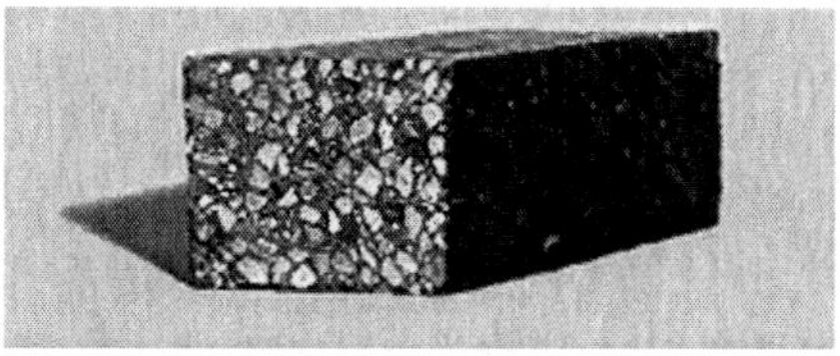

FIGURE 1.12 Sample of RubCon. (From O. Figovsky and D, Beilin, "Building Materials Based on Advanced Polymer Matrix: Review," *J. Scientific Israel Advanced Technology* 10, no. 3 (2008): 1–119.)

More detailed information about structure, properties, and manufacturing of RubCon is presented in Chapter 2.

This material will find wide application as industrial flooring, galvanic and electrolysis baths, supports and foundations, underground structures, high-speed railroad ties, and so on. Application of RubCon in construction allows the builder to solve corrosion problems, negative influence of temperature, degradation of a material at raised UV exposure, radiation, and other adverse natural and technogenic factors. It will allow an increase of the time between repair periods, and in reliability and durability of buildings and structures, especially in aggressive environments.

Polymer Concrete Based on an Organosilicate Matrix [21]

It has been shown that acid-resistant silicate polymer concretes based on liquid glass have high porosity (up to 18% to 20%), low strength and pot life, large shrinkage deformation, and insufficient water resistance. Therefore, they cannot be used as materials for load-bearing structural elements.

A significant increase of silicate matrix strength and density was achieved by incorporation of special liquid organic alkali-soluble silicate additives, which block superficial pores and reduce concrete shrinkage deformation. It was demonstrated that introduction of a tetrafurfuryloxisilane additive dramatically increases strength, durability, and shock resistance of silicate polymer concrete in aggressive media. This effect is attributable to hardening of contacts between silicate binder gel globes and modification of alkaline components owing to "inoculation" of the furan radical. The optimal composition of the pointer concrete was obtained based on an organosilicate matrix (OSPC) with increased strength, chemical resistance in aggressive environments, density, and crack resistance.

Chapter 3 is devoted to a detailed description of the composition and properties of this polymer concrete.

REFERENCES

1. "Intended Use of Polymer Concrete," Systemy i Technologie, http://www.sytec.pl/en/polimerobeton-en.php#wlasciwosci-betonow-zywicznych.
2. Lee Eng Hing. *Application of Polymer in Concrete Construction*. A report submitted in partial fulfillment of the requirements for the degree of Bachelor of Civil Engineering

Faculty of Civil Engineering, Universiti Teknologi, Malaysia, http://www.efka.utm.my/thesis/IMAGES/3PSM/2007/JSB/PARTS7/leeenghingsx031322awj04d07ttt.pdf.

3. Fowler, D. W., "Polymers in Concrete: Where Have We Been and Where Are We Going?," ICPIC 2001, The Tenth International Congress in Polymer Concrete, Honolulu, HI, 2001, USA.
4. Fowler, D. W., "State of the Art in Concrete Polymer Materials in the USA," ICPIC 2007, 12th International Congress in Polymer Concrete, Chancheon, Korea, 2007.
5. Golestaneh, M., Amini, G., Najafpour, G. D., and Beygj M.A., "Evaluation of Mechanical Strength of Epoxy Polymer Concrete with Silica Powder as Filler," *World* Applied *Sciences* Journal 9, no. 2 (2010): 216–220.
6. Blaga, A., and Beaudoin, J. J., *Polymer Concrete*, CBD-242, November 1985 http://web.mit.edu/parmstr/Public/NRCan/CanBldgDigests/cbd242_e.html.
7. Chmyhov, V., "Resistance of Rubber Concrete to Action of Aggressive Environments," doctoral thesis, Voronezh, 2002 (in Russian).
8. Martinez-Barrera, G., Vigueras-Santiago, E., Gencel, O., and Hagg Lobland, H. E., "Polymer Concretes: A Description and Methods for Modification and Improvement," Journal of Materials Education 33, no. 3–2 (2011): 37–52.
9. Garas, V. Y., and Vipulanandan, C., *Review of Polyester Polymer Concrete Properties*, http://www2.egr.uh.edu/~civeb1/CIGMAT/03_poster/11.pdf.
10. Dinitz, A. M., *The Use of Polymer Concrete Materials for Construction, Maintenance, Rehabilitation, and Preservation of Concrete and Steel Orthotropic Bridge Decks*, http://media.brintex.com/Occurrence/21/Brochure/678/brochure.pdf.
11. Ribeiro, M. C. S., Tavares, C. M. L., and Ferreira, A. J. M., "Chemical Resistance of Epoxy and Polyester Polymer Concrete to Acids and Salts," Journal of Polymer Engineering 22, no. 1 (2011): 27–44.
12. Solomalov, V., and Selyaev, V., *Chemical Resistance of Composite Building Materials*. Moscow: Stroyizdat, 1987 (in Russian).
13. Czarnecki, L., "Polymers in Concrete on the Edge of the Millennium," ICPIC 2001, Tenth International Congress in Polymer in Concrete, Honolulu, HI, USA, 2001.
14. Czarnecki, L., *Polymer Concrete*. Warszawa: Arkady, 1982 (in Polish).
15. Borisov, Yu., Potapov, Yu., Figovsky, O., and Beilin, D., "Water Resistance of the Polymer Concretes," *J.* Scientific Israel Advanced Technology 14, no. 3 (2012): 84–91.
16. Moschansky N., and Paturoev V., *Constructional and Chemical Resistance of Polymer Concretes*, Moscow: Stroyizdat, 1970 (in Russian).
17. Potapov, Yu., Borisov, Yu., Barabash D., and Makarova T., *Effective Building Composites Based on Rubber Binders*, Voronezh, 2000 (in Russian).
18. Potapov, Yu., Borisov, Yu., Chmyhov, V. D., and Beilin, D., "Research of Polymer Concrete Based on Low Molecular Polybutadiene, Part VIII: Chemical Resistance of Polymer Concrete," *J. Scientific Israel Technological Advantages* 4, no. 3–4 (2002): 25–31.
19. Figovsky, O., Chmyhov, V., and Beilin, D., "Corrosion Resistance of Rubber Concrete," *Open Corrosion Journal* 3 (2010): 28–37, http://www.benthamscience.com/open/tocorrj/articles/V003/28TOCORRJ.pdf.
20. Figovsky, O., and Beilin, D., "Building Materials Based on Advanced Polymer Matrix: Review," *J. Scientific Israel Advanced Technology* 10, no. 3 (2008): 1–119.
21. Figovsky, O., and Beilin, D., "Optimal Composition, Strength and Chemical Resistance of Silicate Polymer Concrete," ICPIC 2010, 13th International Congress in Polymer in Concrete. Madeira. Portugal, 2010.

2 Polymer Concrete Based on a Vulcanized Polybutadiene Matrix

Development of a manufacturing process for diene oligomers belonging to a liquid rubber class with viscous liquids consistence allowed us to create a new class of conglomerate polymer composite materials—rubber concrete (RubCon). Rubber concrete is an advanced construction material created over the last few years. It is polymer concrete with a unique set of physical–mechanical, chemical, and technological properties that allow the creation of highly effective building structures and products.

RubCon contains no cement as a binder; its matrix is polybutadiene—a polymer from the liquid rubber family—so RubCon has elastic properties and it is extremely resistant to aggressive chemicals, highly repellant to water, and has a remarkable compression strength. It does not exhibit the common failure mechanisms of conventional concrete—such as cracking and flaking, and freeze and thaw—and it resists vibrations, making it an ideal pad material for pumps and compressors. Furthermore, it coats reinforcing rods, making the rods impenetrable by water, and hence arrests corrosion. The strength and durability of concrete is dependent upon the variation of particles and the binder used with its fabrication.

RubCon is applied in the same manner as conventional concrete, formulated first from a component mixture into a liquid and then cured for 12–48 hours for hardening. The initial binder components are formulated off-site into a mixture. The component mixture consists of a single component package for hot curing (150°C to 180°C) with a shelf life of three months (in a closed container), and a two-component package for cold (20°C to 25°C) and semihot (70°C to 100°C) curing with a shelf life of 6 months. The components can be easily formulated on-site in a nontoxic and completely safe manner. RubCon is easily applied and will adhere to metal or glass reinforcements. After 2 days, RubCon may be walked on and after 7 days, it is ready for work loads. With the use of special adhesives, RubCon can be applied over existing concrete flooring. RubCon's outstanding mechanical properties are shown in Table 2.1.

STRUCTURE OF RUBCON [1,2,3]

Selection of the Polymer Part

Compositions based on liquid rubbers are divided, according to their degree of filling, into a number of structural subsystems including some that are described as a

TABLE 2.1
Basic Physical–Chemical and Mechanical Properties of RubCon

Indices	Units	RubCon
Density	kg/m^3	2100–2300
Strength at	MPa	
• compression		80–95
• bending		25–30
• tension		12–15
Modules of elasticity	MPa10^4	2.0–2.7
Poison's ratio		0.26–0.28
Thermal conductivity coefficient	W/m/°C	0.3–0.5
Wear resistance	$(kg/m^2)10^{-3}$	2–3
Specific toughness	$(J/m^2)10^3$	3.5–4.5
Heat stability	°C	80–100
Water absorption	%	0.05–0.06
Coefficient of chemical resistance at 20°C (based on 360 days of exposure)		
• 20% H_2SO_4		0.97–0.98
• 10% Lactic acid		0.95–0.96
• 20% Caustic potash		0.97–0.98
• 35% H_3PO_4		0.96–0.98
• Water		0.99–0.995
• Saltwater		1.00–1.05
Resistance to abrasion	$(kg/m^2)10^{-3}$	2–3.5

Source: Reprinted from O. Figovsky and D. Beilin, "Building Materials Based on Advanced Polymer Matrix: Review," *J. Scientific Israel Technological Advantages* 10, no. 3 (2008): 1–119. With permission.

matrix + additives: liquid rubber + hardening components, rubber matrix (RM), RM + filler, rubber binder (RB), RB + complementary additives, and rubber concrete (RubCon).

A very important time in the creation of a network polymeric composition is the choice of oligomer, because its chemical composition and structure determine characteristics of the created material. This is true also for RubCon, the liquid phase of which consists from rubbers with various microstructures of polymeric chains. Liquid rubbers in projected compositions are capable, if acted on by special sulfur-accelerating systems, to be vulcanized with formation of space-linked net polymers, the space net of which mainly determines the positive properties of the hard base of the RubCon composite.

As several studies show, the best combination of physical, chemical, and technical parameters is exhibited by RubCon, based on diene oligomers without functional groups hardened in the presence of a sulfur-acceleration system. Composites of this kind have a number of positive properties, and have excellent characteristics of durability, and crack, water, and chemical resistance.

For the correct selection of rubber for RM, a complex of the following characteristics needs to be taken into account, when considering polymers that satisfy the properties needed for the projected composite, its availability, and technologic suitability. Following these criteria, the selection of the polymer was carried out among existing kinds of liquid rubber: SKDN-N® (Russia), PBN® (Russia), Polyoil 110® (Germany), and Ricon 130® (USA). All of these liquid rubbers have low molecular hydrocarbon diene-based polymers (solvent-free liquid oligomers of butadiene) with a large spectrum of molecular masses and microstructures. These are characterized by numerous nonsaturated olefin bonds allowing easy linkage and chemical modifications. Moreover, each of them has a determined and carefully controlled microstructure, which seriously determines the properties of liquid rubber. According to the microstructure of the polymeric group, the selected types are oligomers with 1, 4-cis structure (SKDN-N and Polyoil 110) and rubbers with mixed microstructure (PBN and Ricon 130).

In order to select the liquid rubber to use in the mixture, the durability of considered kind of RubCon was studied. For this goal, prism samples sized 40 × 40 × 160 mm were tested.

According to the results of the experiment (Figure 2.1), the maximum compression and bending stresses were found for samples based on the liquid rubber PBN (σ_{com} = 93.0 MPa, σ_b = 28.0 MPa) and Ricon 130 (σ_{com} = 94.0 MPa, σ_b = 26.0 MPa). Less durability was found for samples based on Polyoil 110 (σ_{com} = 84.0 MPa, σ_b = 23.5 MPa) and SKDN-N (σ_{com} = 81.00 MPa, σ_b = 22.0 MPa).

Upper values of the durability of RubCon samples based on PBN and Ricon 130 in comparison with those of SKDN-N and Polyoil 110 are explained, in our opinion, by different molecular structures of polymeric chains of the used oligomers. As mentioned previously, rubbers SKDN-N and Polyoil 110 contain a major amount of 1,4-cis units (up to 75% of the total), while in rubbers with a mixed microstructure of PBN and Ricon 130, the unit content of 1,4-cis, 1,4-trans, and 1,2

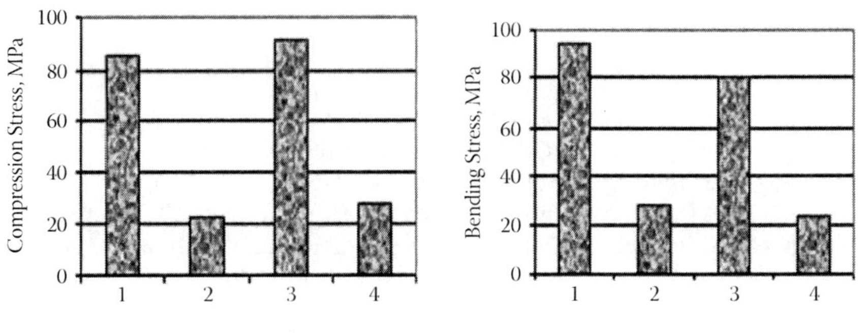

FIGURE 2.1 Values of σ_{com} and σ_b of RubCon for various rubbers. (I) Polyoil®, (2) PBN®, (3) Ricon 130®, (4) SKDN-N®. (Reprinted from O. Figovsky and D. Beilin, "Building Materials Based on Advanced Polymer Matrix: Review," *J. Scientific Israel Technological Advantages* 10, no. 3 (2008): 1–119. With permission.)

is comparable. Such a redistribution of the microstructure of polymeric chains is favorable for increasing the three-dimensional polymer induction formation period and the degree of double bond conversion. Moreover, rubbers with a mixed microstructure have better thixotropy, which allows better mixing of the components when the RubCon mixture is prepared. These factors are favorable for the processes that take place in the compositions when they are prepared and vulcanized, and cause better durability characteristics of the product. However, we must note that diene oligomers of the 1,4-cis structure are more reactionable, and based on them, RubCon had a lower temperature and a higher rate of vulcanization with other comparable data.

Linear polybutadiene oligomers of a mixed microstructure of PBN and Ricon 130 was found to be optimal as the basic polymer for RubCon.

Optimization of Rubber Binder Viscosity

Since viscosity is a very important parameter characterizing every liquid rubber and influencing the main physical and mechanical properties of the projected composite, it frequently determines the choice of the polymer and reasonability of its use. The industry manufactures rubbers with viscosity varied in large spectrum (e.g., for polybutadiene PBN, that interval is [(0.2 ÷ 7.1) Pa · s]). However, there is no data that allows estimation of the viscosity of rubber that is useful in RubCon compositions.

Obviously, the lower the viscosity of the rubber, the lower the viscosity of the obtained RubCon composition; hence, there is opportunity to obtain better mixing of the components and so improve physical and mechanical properties of the composite or reduce the consumption of the polymer. On the other hand, low-viscosity rubber that is caused by a large presence of molecules with a polymerization degree from 3 to 6 cannot provide vulcanization net of high density, which decreases the durability of the composite in general. Although, use of rubber with viscosity that allows formation of vulcanization net of high density meant excellent physical and mechanical characteristics of the composite may cause increased o viscosity of the entire composition. The last factor would cause increased consumption of energy and effort for the preparation of the mixture without gaining better properties, or make the mixture loose and bad for treatment, with worse resulting properties.

A series of experiments were carried out to evaluate the value of the viscosity of rubber allowing the needed varied parameter has been viscosity, and the measured function — the *RubCon* durability against compression and bending. The experiment used PBN rubber with viscosities (Pa · s) of 0.4, 0.6, 0.8, 1.0, 1.2, 1.4, 1.6, 1.8, 2.0, and 2.2. The regression analysis of the obtained results allowed deduction of mathematical models that adequately describe the correlation of the durability of the samples to compression and bending:

$$\sigma_{com} = 46.377\eta(1 - 0.36\eta) + 60.731 \tag{2.1}$$

$$\sigma_b = 33.564\eta(1 - 4.2\eta) + 10.553 \tag{2.2}$$

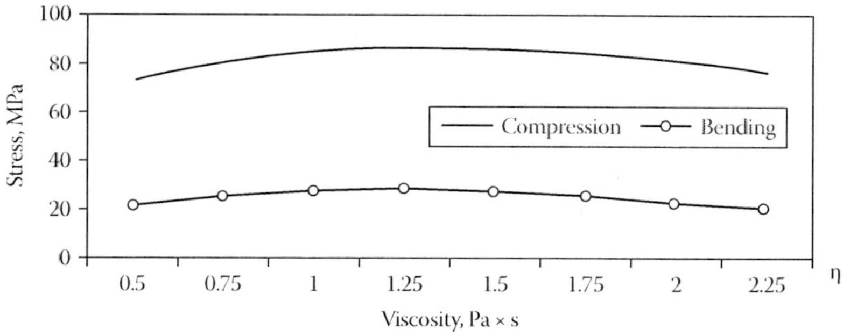

FIGURE 2.2 Stresses σ_{com} and σ_b of RubCon depend on amount of rubber viscosity. (Reprinted from O. Figovsky and D. Beilin, "Building Materials Based on Advanced Polymer Matrix: Review," *J. Scientific Israel Technological Advantages* 10, no. 3 (2008): 1–119. With permission.)

where σ_{com} is the durability of PBN RubCon samples against compression, MPa; σ_b is the durability against bending; η is the rubber viscosity, Pa · s.

The obtained correlation is described by parabolic curves 1 and 2 (Figure 2.2) characterized by initial rising with further extreme value, then decrease of the function. Maximal values of the curves are obtained at different values of viscosity. For curve 1 of the change of durability against compression, the maximum is obtained at viscosity 1.2 Pa · s.

This is probably explained by the hardness of RubCon prepared from rubber with viscosity 1.6 Pa · s in comparison with that from rubber with viscosity 1.2 Pa · s, because the density of linkage is higher for a polymer with higher molecular mass related to viscosity.

Based on these results, we can conclude that the value of optimal viscosity of low-molecular polybutadiene with mixed microstructure, involved in the RubCon mixture, is in an interval from 1.1 to 1.7 Pa · s. The further studies employed rubber with viscosity 1.5 Pa · s.

Estimation of Optimal Rubber Content

It is obvious that the amount of low-molecular rubber in the RubCon composition permanently changes with the kind, dispersion, and amount of the filler; number and granular composition of the fillers, and so on. However, the evaluation of the optimal content of rubber in the composition is possible at the initial stage of the composition projecting. To solve this problem, an experiment was carried out. The variable quantity was the amount of low-molecular rubber in the composition (PBN); the goal functions were the compression and bending strength of RubCon. The composition and technology of sample preparation have been the same as that in selection of the kind of rubber. A regression analysis of the results (Figure 2.3) provided adequate mathematical models:

$$\sigma_{com} = 45.4\ \lambda\ (1 - 0.052\ \lambda) + 116.9 \tag{2.3}$$

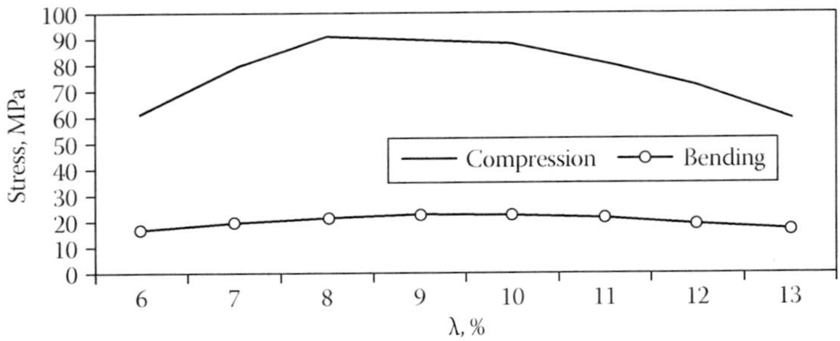

FIGURE 2.3 Stresses σ_{com} and σ_b of *RubCon* depend on amount of rubber. (Reprinted from O. Figovsky and D. Beilin, "Building Materials Based on Advanced Polymer Matrix: Review," *J. Scientific Israel Technological Advantages* 10, no. 3 (2008): 1–119. With permission.)

$$\sigma_b = 18.4\,\lambda\,(1 - 0.052\,\lambda) - 58.2 \tag{2.4}$$

where λ is the content of rubber.

The analysis of the obtained data shows that the optimal content of rubber in the composition is from 8.5 to 10.5 mass%. Rubber content of less than 8.5 mass% causes discontinuance of the film structure of the polymeric matrix and pore formation, while more than 10.5 mass% provides an unstable inhomogeneous structure of the mixture during formation.

Choice of Hardening Agent for Rubber Matrix

Liquid polybutadienes without functional groups may be vulcanized on double bonds of the diene part of the polymeric chain in the presence of a sulfur-accelerating, redox, or peroxide system. However, only the sulfur-accelerating system is able to provide the maximal durability values. Sulfur also has other advantages such as low price, availability, and so on. The amount of involved sulfur in the system depends on the desired properties of the product. For hard RubCon, this is 47–55 mass parts per 100 mass parts of rubber.

The rising of the rate of chemical reactions between sulfur and rubber, acceleration of the vulcanization, and reduction of its temperature are accomplished using special accelerators. The use of a few accelerators may improve the vulcanization process because of their mutual activation. Two accelerators were used: tetramethylthiuram-disulfide (thiuram-D) and CAPTAX®. They act mutually and thus improve the vulcanization process and the resulting properties of RubCon.

The previously mentioned suggestion has been tried in a two-phase experiment in which the variable parameters were the amounts of accelerators, and the value of RubCon durability as the function.

It has been found from experience that the maximal value of RubCon durability is obtained at the CAPTAX content of 0.05 mass%, and thiuram-D 0.35 mass%. The maximum durability against compression has been found at a maximum content of thiuram-D and moderate content of CAPTAX.

Regression analysis of the experiment results provides the following empirical equations:

1. At a thiuram-D content of 0.15 mass%:

$$\sigma_{com} = 1120\, k\, (l - 8.9k) + 65.1 \quad (2.5)$$

2. At a thiuram-D content of 0.25 mass%:

$$\sigma_{com} = 715\, k\, (l - 9.3k) + 78.213 \quad (2.6)$$

3. At a thiuram-D content of 0.35 mass%:

$$\sigma_{com} = 510\, k\, (l - 8.8k) + 88.25 \quad (2.7)$$

where k is the mass content of CAPTAX (%).

4. At a CAPTAX content of 0.03 mass%:

$$\sigma_{com} = 4\, t\, (l + 22.5t) + 87.075 \quad (2.8)$$

5. At a CAPTAX content of 0.05 mass%:

$$\sigma_{com} = 190\, t\, (t - 0.49) + 101.28 \quad (2.9)$$

6. At a CAPTAX content of 0.07 mass%:

$$\sigma_{com} = 240\, t\, (t - 0.35) + 101.25 \quad (2.10)$$

where t is the mass content of thiuram-D (%).

The general correlation equation:

$$\sigma_{com} = 173.3t\, (t - 0.186) - 7042k\, (k - 0.122) + 300tk + 73.28 \quad (2.11)$$

The decrease of RubCon durability at a CAPTAX content of less than 0.05 mass% is caused by the violation of accelerators' complex action and appearance of chemically free rubber. Content values of more than 0.05% causes excess CAPTAX and a related decrease in durability. The increase of product durability, when the content of thiuram-D approaches 0.45% is caused by gas deliverance and pore formation in the composite volume.

The hypothesis of a positive influence of CAPTAX on durability because of the kinetic factor has been checked in the following experiment.

A series of RubCon samples were prepared. The content of the main components (except accelerators) was the same. The first composition used both thiuram-D

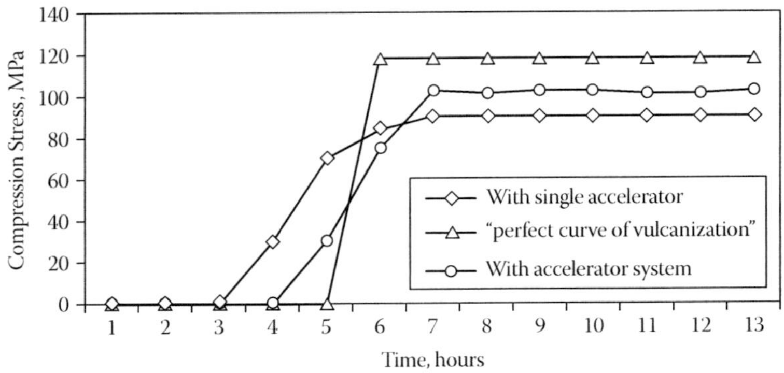

FIGURE 2.4 Kinetic curves of: (1) with single accelerator, (2) with accelerator system, (3) "perfect curve" of vulcanization. (Reprinted from O. Figovsky and D. Beilin, "Building Materials Based on Advanced Polymer Matrix: Review," *J. Scientific Israel Technological Advantages* 10, no. 3 (2008): 1–119. With permission.)

and CAPTAX accelerators; the second composition used thiuram-D only. After the temperature reached 120°C in exothermal regime, each 60-mm three-prism sample (40 × 40 × 160 mm) was removed from the camera and tested against compression.

The analysis of curves 1 and 2 (Figure 2.4) shows that involving the second accelerator in the composition improves the kinetics of vulcanization for the following reasons:

- An increase in the induction period of vulcanization (AB) when the structure is formed;
- A reduction of the main period of vulcanization when the optimum is gained;
- Increase in the duration of the optimum.

The change of parameters of vulcanization kinetics found in the experiment allowed more order in the composite structure, a reduction of the number of dislocations, and consequently, improvement of the mechanical characteristics of RubCon. Let us note that the optimum content of the components in the system with the additive action of accelerators for rubber of the same kind is constant; for PBN it is 7 parts thiuram-D per 1 part CAPTAX.

Planning of Activator and Calcium-Containing Components of the Composition Rubber Matrix—RubCon

Based on an analysis of references, it was found that organic vulcanization accelerators are especially active in the presence of several oxides and hydroxides of metals (vulcanization activators) like zinc, lead, magnesium, calcium, cadmium, bismuth, and their combinations.

The most widespread activator used in rubber composition treatment is zinc oxide (zinc white). In comparison with other activators, it is cheaper and largely used in the chemical industry as raw material.

Zinc oxide was accepted as an RM vulcanization activator, and its optimal content in the RubCon composition was estimated at 10–20 mass parts per 100 mass parts of rubber.

Use of 0.5 mass% calcium-containing component (CaO) allows reduction of gas deliverance and pore formation during vulcanization.

Optimization of the Rubber Matrix Composition: RubCon for Strength

The evaluation of RM optimal composition was carried out by varying of three parameters: the amount of sulfur (s), the amount of accelerator (t), and the amount of the activator (z) in RM, whereas the efficiency functions are compression and bending strength.

The amount of PBN rubber in the experiment was 100 mass parts; the content of other components was CaO (5 mass parts), filler (87.5 mass parts), sand (300 mass parts), and coarse aggregate (680 mass parts).

The regression analysis of results of the experiment provided the following equations:

1. For samples tested against compression:

$$\sigma_{com} = 0.34\, s\,(s - 100) - 0.28\, t\,(t - 23.79) - 0.14\, z\,(z - 8.36) - 782.1 - 0.08st - 0.06sz - 0.01tz \tag{2.12}$$

2. Those tested against bending:

$$\sigma_{b} = 0.07\, s\,(s - 100) - 0.13\, t\,(t - 17.07) - 0.06\, z\,(z - 59.2) - 197.9 - 0.02st - 0.02sz - 0.06tz \tag{2.13}$$

The results of the experiment showed that sulfur and vulcanization activators had the most influence on the change of durability, while that of the accelerator system is less important. The increase of sulfur content increases durability against compression, but decreases durability against bending. The interactions of the sulfur accelerator and sulfur activator have similar influence.

The compositions of RubCon designed for operations under compression or bending, respectively, are given in Table 2.2.

TABLE 2.2
Optimal Composition RM for RubCon—PBN®

	Mass Parts of Components					
Tests	**Rubber**	**Sulfur**	**Thiuram-D®**	**Captax®**	**Zinc Oxide**	**Calcium Oxide**
Compression	100	50.0	4.4	0.7	15.6	6.3
Bending	100	47.5	4.3	0.5	18.8	5.6

Source: Reprinted from O. Figovsky and D. Beilin, "Building Materials Based on Advanced Polymer Matrix: Review," *J. Scientific Israel Technological Advantages* 10, no. 3 (2008): 1–119. With permission.

Control tests of RubCon samples prepared according to the foregoing recommendations provided the following strength values: 105 MPa at compression, 31 MPa at bending.

PHYSICAL–MECHANICAL PROPERTIES OF RUBCON [1, 4–8]

Stress–Strain Relationship at Central Uniaxial Compression

The experimental research of RubCon deformation at uniaxial central compression was carried out. A special testing module consisting of three calibrated rods provided constant deformation speed (Figure 2.5).

The characteristic pressure–deformation relationship is shown in Figure 2.6.

The analytical form of this dependence may be written as:

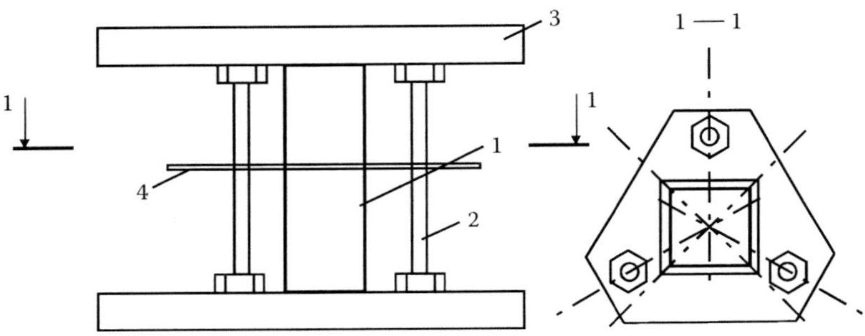

FIGURE 2.5 Draft of testing gadget: (1) sample of RubCon, (2) calibrated rods, (3) plate, (4) centering bar. (Reprinted from Yu. Potapov, O. Figovsky, Yu. Borisov, S. Pinaev, and D. Beilin, "Stress–Strain State of Compressed Elements from Polymer Concrete," *J. Scientific Israel Technological Advantages* 4, no. 3–4 (2002): 20–24. With permission.)

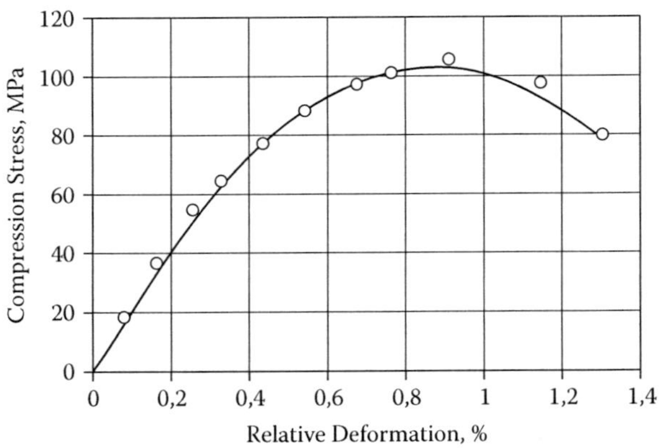

FIGURE 2.6 Relationship between relative deformation of RubCon and compressive stress. (Reprinted from Yu. Potapov, O. Figovsky, Yu. Borisov, S. Pinaev, and D. Beilin, "Stress–Strain State of Compressed Elements from Polymer Concrete," *J. Scientific Israel Technological Advantages* 4, nos. 3–4 (2002): 20–24. With permission.)

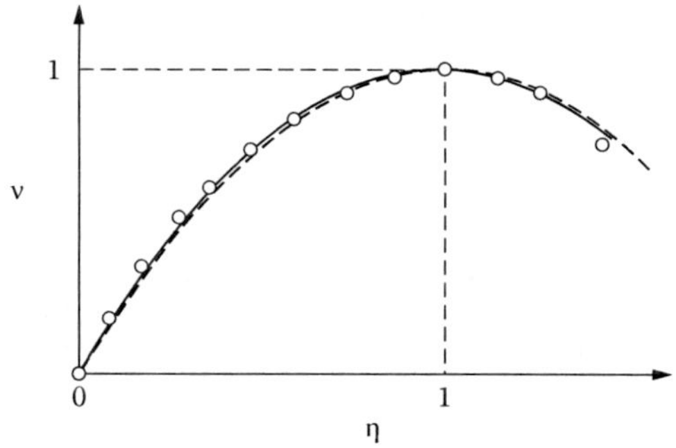

FIGURE 2.7 Diagram of function (Eq. 2.16). (Reprinted from Yu. Potapov, O. Figovsky, Yu. Borisov, S. Pinaev, and D. Beilin, "Stress–Strain State of Compressed Elements from Polymer Concrete," *J. Scientific Israel Technological Advantages* 4, nos. 3–4 (2002): 20–24. With permission.)

$$\sigma = 2.6 + 241.4\varepsilon - 146\varepsilon^2. \quad (2.14)$$

Let $\nu = \sigma/\sigma_R$; $\eta = \varepsilon/\varepsilon_R$; $k = E\varepsilon_R/\sigma_R$. On the experimental base it is believed that dependence $\nu = f(k\eta)$ looks like a square parabola $\nu = k\eta - \eta^2$, where σ, ε are the current values of compression stress and deformations, respectively, σ_R, ε_R are the coordinates of the diagram top, k is the factor describing the elastic–plastic properties of RubCon. Then,

$$\sigma = \sigma_R\,[k(\varepsilon/\varepsilon_R) - (\varepsilon/\varepsilon_R)^2] \quad (2.15)$$

and

$$\eta = 0.5[k \pm (k^2 - 4\,\nu)^{0.5}]. \quad (2.16)$$

The diagram of Function (2.16) (dashed line) in relative coordinates coincides well with experimental dependence (a continuous line) stress–strain (Figure 2.7).

Temperature Effect

For the experiment, test RubCon samples sized 40 × 40 × 160 mm were prepared. Tests were carried out in the special chamber, in the temperature range −80°C ± +80°C appropriate to real operating conditions of the material. During experiments, the stress–strain state of samples was determined depending on the temperature of the environment. In particular, the changes of the modulus of elasticity, ultimate strength at compression, and the ultimate deformations of a material at influence of temperature were determined in comparison with similar values obtained in testing of control samples at room temperature.

Results of experiments at a negative temperature range are illustrated in Figure 2.8. It is possible to see that at the negative temperature (−80°C), the ultimate relative strain decreases on the average of 14% in comparison with control values, and ultimate strength at compression and the modulus of elasticity, on the contrary,

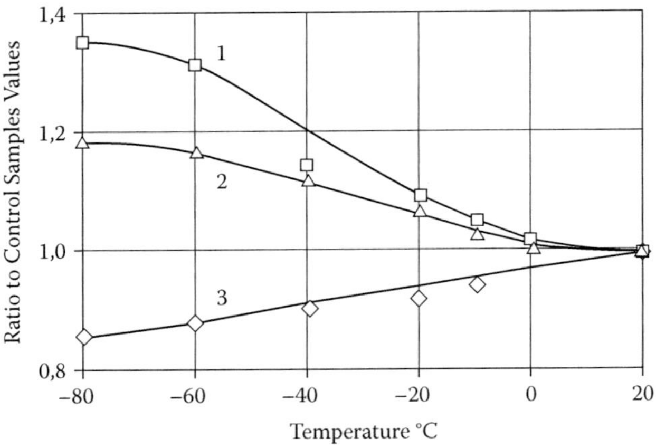

FIGURE 2.8 Influence of *negative* temperatures on the ratio of the module of elasticity (1), ultimate compression strength (2), and ultimate deformations at compression (3) to the similar values obtained at test of control RubCon samples at room temperature. (Reprinted from Yu. Potapov, O. Figovsky, Yu. Borisov, V. Chmyhov, and D. Beilin, "Influence of Temperature on Physical–Mechanical Characteristics of Polymer Concrete," *J. Scientific Israel Technological Advantages* 5, nos. 1–2 (2003): 11–13.)

are increased by 19% and 35%, respectively. It is necessary to note that linear changes of ultimate deformation depend on temperature.

Researchers [6] showed that the microstructure of RubCon has elastic, elastic–plastic, and viscous phases. The amount of elastic in the composite is less in comparison with others, and consequently deformability of RubCon at action a long-term and a short-term loadings in the greater degree is determined by elastic and elastic–plastic deformations. The increase of the strength of RubCon and the modulus of elasticity at compression and the decrease of its ultimate deformations at negative temperatures can be explained by increases in viscous phase viscosity and partial transformation of an elastic–plastic phase of a composite in elastic. The increase in the elastic phase results in embrittlement of the composite but no changes in its stress–strain state.

The influence of increased positive temperatures in the range 20°C–80°C on RubCon mechanical characteristics is shown in Figure 2.9.

An increase in temperature from 20°C up to 50°C has a small effect on RubCon. At a further increase in temperature, it was possible to expect substantial growth of ultimate deformations along with a simultaneous drastic decrease in ultimate compressive strength and modulus of elasticity. At the maximum positive temperature of 80°C, deformability of RubCon increases 2.5 times, ultimate strength is reduced by 40%, the modulus of elasticity decreases by 50%. At rise in temperature the part of an elastic–plastic phase transfers to viscous; thus, highly elastic and viscous deformations develop and produce increases in RubCon deformability, and decreases in its ultimate compressive strength and the module of elasticity.

These deformations can be reversed, that is, if temperatures are decreased, there is a return process. The effect of temperature depends on its value and the duration

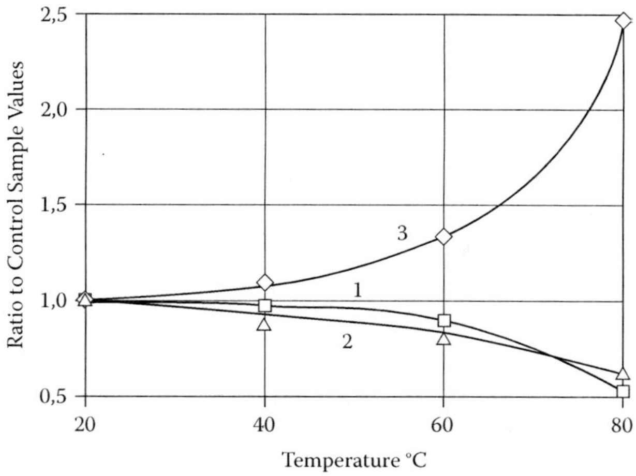

FIGURE 2.9 Influence of *positive* temperatures on the ratio of the module of elasticity (1), ultimate compression strength (2), and ultimate deformations at compression (3) to the similar values obtained at test of control *RubCon* samples at room temperature. (Reprinted from Yu. Potapov, O. Figovsky, Yu. Borisov, V. Chmyhov, and D. Beilin, "Influence of Temperature on Physical–Mechanical Characteristics of Polymer Concrete," *J. Scientific Israel Technological Advantages* 5, nos. 1–2 (2003): 11–13. With permission.)

of heating or cooling. Short-term heating causes the convertible weakening of composite structure as a result of temperature binder plasticization. The result of this process is decreased RubCon compressive strength and modulus of elasticity and an increase in deformability.

Thermomechanical Destruction of RubCon

As we know, the load-carrying capacity of any structural material depends on the relationship of external force factors, temperature, and the duration of their action on strength, thermostability, and durability. This means that the problem of forecasting and increasing the material's serviceability is reduced to determining the interrelations between these material behavior characteristics. Thermomechanical destruction is one of most practically important kinds of failure in high molecular composite materials, along with breaks of chemical connections. Stresses at the point of deformation grow under loading and the probability of change in macromolecular chains increases as well. These chains are oriented in a direction of external force action, which determines the elastic deformation of molecules whose changes of position may be responsible for their destruction [3,5].

Based on the conventional concept of S. Jurkov [7] that mechanical failure of a polymeric material is a process its thermochemical destruction, V. Jartsev [8] put forward the formula of the load-carrying capacity *time border* of a composite material based on its strength and an external temperature field.

$$\tau_f = \tau_m * \exp\,[(U_o - \gamma * \sigma)(R * T)(1 - T/T_m)] \tag{2.17}$$

where τ_f, T, σ is the time to failure (sec), temperature (°K), and strength (MPa), respectively; τ_m is the minimum durability of the material (sec), U_o is the activation energy of a failure (thermomechanical destruction); γ is the coefficient of external mechanical forces; R is the universal gas constant; and T_m is the ultimate design temperature for the material.

Two other borders of material load-carrying capacity follow from Equation (2.17).

The *ultimate strength*, depending on temperature and the duration of its action:

$$\sigma = (1/\gamma)*\{U_o - [2.3* R* T* lg(\tau_f/\tau_m)/1 - T/T_m]\} \tag{2.18}$$

The *ultimate temperature*, depending on strength and time:

$$T = (1/T_m) + [2.3* R* lg(\tau_f/\tau_m)/ U_o - \gamma*\sigma] \tag{2.19}$$

Formulas (2.17) and (2.18) describe the interrelation of the stress–strain state of the composite material and duration of temperature and external loading actions. Thus, for an estimation of load-carrying capacity of a composite material and its directed regulation, it is necessary to know all of the material's physical constants. For this purpose, we have undertaken experimental research at a number of preset values of the normal stresses and temperature durations of sample destruction.

Prismatic beam samples (30 × 60 × 700 mm) of RubCon were tested for a cross, so-called *pure* bend, by two equal concentrated forces symmetrically located at an average third of the samples' span and simultaneous temperature influence. Loading was carried out step by step to the attainment of the normal stress in the middle span of 5.7, 6.1, and 6.5 MPa.

For creation of increased temperatures, special spiral heating elements were placed at the sides of the sample. The level of temperature heating was 60°C–105°C. During the test, loading was done after a steady temperature was attained. Temperature was controlled in a maximum stress zone.

The results of the experiment are shown in Table 2.3. The load-carrying capacity values are averaged for test results for various stresses and temperature levels.

The experimental data described by Formula (2.17) are shown in Figure 2.10. Each line that converges several isobar families corresponds to the given level of normal stress σ_1; σ_2; σ_3. Values for T_m and τ_m can be found from Figure 2.10. Two constants, U_o and γ, can be determined from an inclination of straight lines by the formula of activation energy:

$$U(\sigma) = 2.3*R*\Delta lg\ \tau(\sigma)/\Delta 1/T \tag{2.20}$$

For each given loading value, it is possible to determine $U(\sigma)$ (see Figure 2.11).

It can be seen that $\gamma = tg\ \alpha$ and U_o is an ordinate of the straight line $U - \sigma$ at the extrapolation to $\sigma = 0$.

TABLE 2.3
Experimental Values of Stress, Temperature, and Time to Failure

Normal Stress, σ(MPa)	Temperature, T (°C)	Time to Failure, τ_f (min)
$\sigma_1 = 5.65$	72	527
	77	340
	78	132
	86	83
	105	3
$\sigma_2 = 6.07$	65	552
	66	340
	71	215
	72	74
	76	56
	84	42
$\sigma_3 = 6.47$	60	300
	64	265
	67	151
	78	21

Source: Reprinted from O. Figovsky, Yu. Potapov, T. Makarova, and D. Beilin, "Load-Carrying Capacity of Polymer Concrete with Polybutadiene Matrix," *J. Scientific Israel Technological Advantages* 4, no. 1–2 (2002): 21–24. With permission.

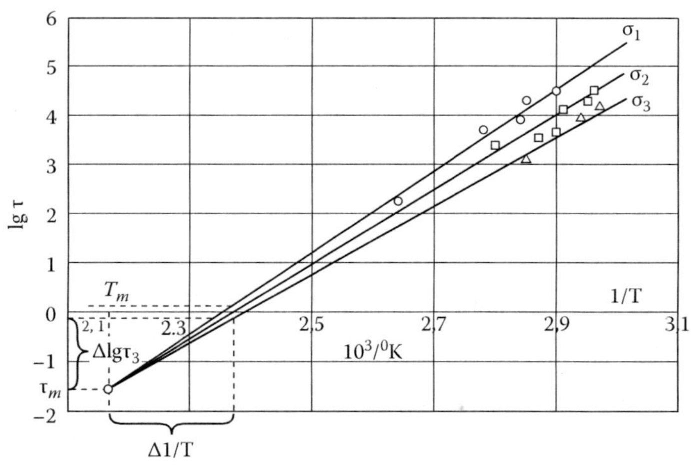

FIGURE 2.10 Test results for different levels of stresses and temperatures. (Reprinted from Figovsky, O., Potapov, Yu., Makarova, T., and Beilin, D. "Load-Carrying Capacity of Polymer Concrete with Polybutadiene Matrix," *J. Scientific Israel Technological Advantages* 4, no. 1–2 (2002): 21–24. With permission.)

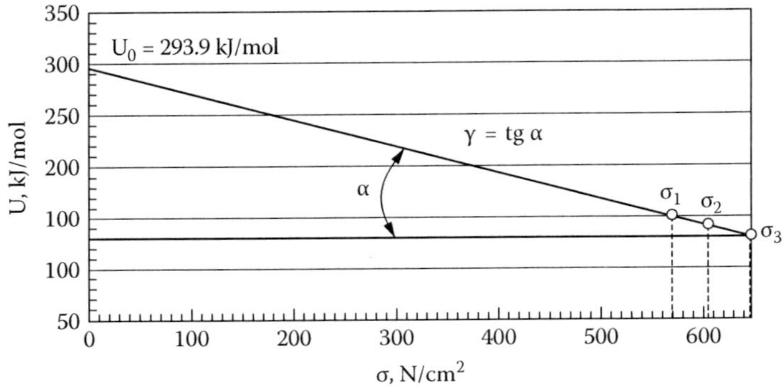

FIGURE 2.11 Relationship activation energy U and normal bending stress σ. (Reprinted from Figovsky, O., Potapov, Yu., Makarova, T., and Beilin, D. "Load-Carrying Capacity of Polymer Concrete with Polybutadiene Matrix," *J. Scientific Israel Technological Advantages* 4, no. 1–2 (2002): 21–24. With permission.)

TABLE 2.4
Constants and Load-Carrying Capacity Parameters of RubCon

Load-Carrying Capacity Parameters					
U_o (kJ/mol)	τ_m (sec)	T_m (10^3/°K)	τ_f (sec)	σ (MPa)	T (°K)
293.9	0.025	2.15	$10^{4.1}$	5.65	352

Source: Reprinted from O. Figovsky, Yu. Potapov, T. Makarova, and D. Beilin, "Load-Carrying Capacity of Polymer Concrete with Polybutadiene Matrix," *J. Scientific Israel Technological Advantages* 4, no. 1–2 (2002): 21–24. With permission.

Values of all constants calculated on the experimental database are given in Table 2.4. From Table 2.4, it follows that the value of U_o is close to the energy at which chemical connections break, resulting in the formation of free macroradicals [9]. The temperature T_m corresponds to the initial temperature of rubber decomposition. The value of a structural-mechanical constant γ corresponds to complex composite materials on the rubber and phenol-formaldehyde resin base. It should be noted that experimentally determined minimum time of destruction, τ_m, is in 100 times exceeds the period of atoms' fluctuation in a solid body. Such a sharp increase τ_m is obviously connected to a significant quantity of filler in the RubCon material (about 90%), which results in a complex manner of crack development at destruction [8].

The resulting values of constants allow determination of the key parameters of load-carrying capacity of the researched composition in the given range of loadings, temperatures, and operation time.

Obviously, to increase the load-carrying capacity of a material, it is necessary to increase the constants T_m, τ_m, and U_o and to reduce γ. It is known [8], that the increase in U_o is promoted by an increased filling and its uniformity. Introduction of special additives (plasticizers, antioxidants, and others) to a composition, along with the manufacturing process and material structure, will have an essential influence on constants T_m, τ_m, and U_o.

It is also known [8] that for various kinds of plastics, the ratio between design-bending stress and tension stress is 0.5–0.87, that is, $\gamma_{bend}/\gamma_{tens} \approx 0.5–0.87$. The ratio between design compression and tension stresses is $\gamma_{compr}/\gamma_{tens} \approx 0.95$.

Knowledge of these relationships and constants will allow forecasting of the limits of RubCon load-carrying capacity, and by their directed regulation, increase its durability.

Summary

- Experimentally obtained temperature coefficients of strength and deformation characteristics of the material will allow design of load-bearing and protecting structures in an operational temperatures range.
- The ratio between design compression and tension stresses is $\gamma_{compr}/\gamma_{tens} \approx 0.95$. This will allow forecasting of the limits of RubCon load-carrying capacity and durability of RubCon.

STRENGTH OF REINFORCED RUBCON [1,3–6,10–17]

Combined Reinforcement and Polymer Concrete Matrix

The combination of steel reinforcement and a composite matrix depends on the following:

- Coefficients of linear expansion of polymer concrete and steel reinforcements
- The friction forces related to reinforcing draft in the process of composite hardening and shrinkage
- Mechanical bond of a reinforcement and a matrix
- Adhesion of a metal reinforcement to a matrix

Coefficient of Thermal Linear Expansion of RubCon

It has been known that the average coefficient of thermal linear expansion is equal to the relation of a sample length increment Δl in temperature interval $[t_1, t_2]$ to the size of this interval $t_1–t_2$. The average coefficient of linear expansion of RubCon was determined with a thermostat unit (Figure 2.12).

Heating of a RubCon sample was done with speed not exceeding 1.5° ± 0.5°C per minute in a range.

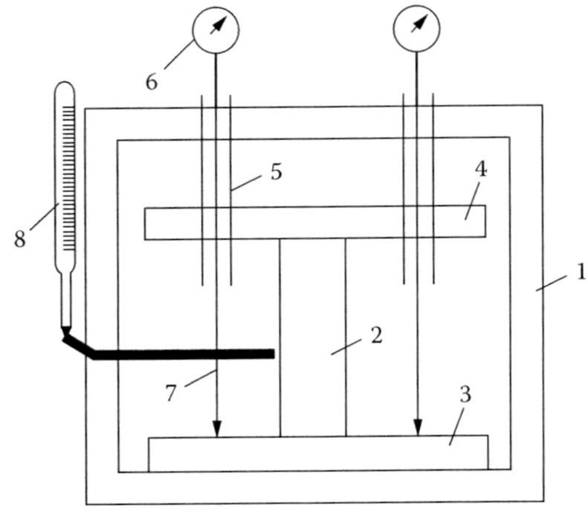

FIGURE 2.12 Test unit for determination of coefficient linear expansion of RubCon: (1) temperature chamber, (2) RubCon sample, (3, 4) steel plates, (5) pile, (6) indicator, (7) leg of indicator, (8) thermometer. (Reprinted from Yu. Potapov, O. Figovsky, Yu. Borisov, S. Pinaev, and D. Beilin, "Joint Work of Reinforcement and Polymer Concrete Matrix," *J. Scientific Israel Technological Advantages* 4, nos. 3–4 (2002): 14–19. With permission.)

$$t_1 = 23°C \div t_2 = 120°C$$

Measurements of the elongation of test samples of the size 40 × 40 × 160 mm during heating were carried out in a thermostatic mode. The average coefficient of linear thermal expansion α was calculated with the following formula:

$$\alpha = 1/l_0\ (\Delta l/\Delta t) \tag{2.21}$$

where l_0 is the initial length in mm of a sample measured at temperature 23°C; Δl is the increment of length of a sample in mm at the temperature interval borders $[t_1, t_2]$, $\Delta t = t_2 - t_1$.

As a result of this research it is established that for RubCon samples,

$$\alpha = 1.35 * 10^{-5}\ (1/°C) \tag{2.22}$$

We shall note that the coefficient of linear expansion of steel is $\alpha = 1.2 \div 1.3 * 10^{-5}\ (1/°C)$ at the same temperature range. Thus, in RubCon-reinforced building structures, there will be no additional internal temperature stresses in all ranges of operational temperature.

Shrinkage Stress of the RubCon Matrix

Dependence of internal shrinkage stresses of the RubCon matrix at the time of its manufacture was determined with the help of a special device (Figure 2.13).

A RubCon layer thickness of 10 mm was put on the surface of a steel console plate measuring 350 × 20 × 2 mm. Then the device was placed in a drying cabinet

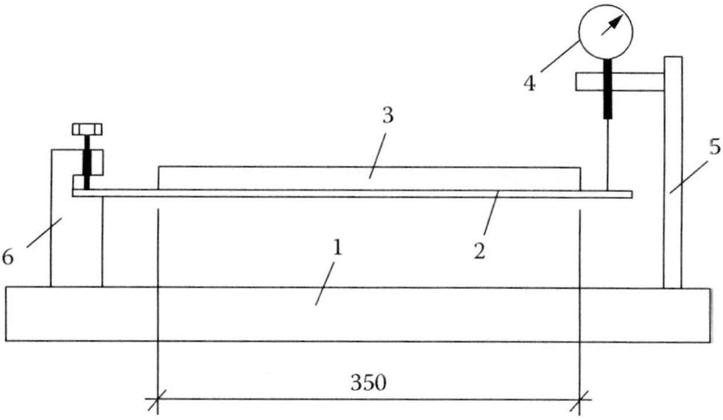

FIGURE 2.13 Test device for determination of shrinkage stress of RubCon matrix: (1) rigid frame, (2) steel substrate, (3) layer of RubCon, (4) indicator, (5) stand, (6) holder. (Reprinted from Yu. Potapov, O. Figovsky, Yu. Borisov, S. Pinaev, and D. Beilin, "Joint Work of Reinforcement and Polymer Concrete Matrix," *J. Scientific Israel Technological Advantages* 4, nos. 3–4 (2002): 14–19. With permission.)

at 120°C, where the RubCon was hardened. By means of the dial-type indicator, displacements of the free end of a console substrate were fixed during RubCon vulcanization. Initial indication was made after 2 hours of isothermal exposure of the sample, that is, at the moment when the vulcanization began and further at regular time intervals.

Displacement of the free end of a steel plate substrate occurs due to shrinkage stresses in a RubCon layer. Knowing the value of this displacement, it is possible to determine shrinkage stresses with the formula:

$$\sigma_{sh} = \{4\Delta * E_s * k\, [I_s + I_{RC}\,(E_{RC}/E_s)]\}/F_{RC} * l_{RC}^2 * (t_s + t_{RC}) \tag{2.23}$$

where Δ is the displacement of free end of steel console substrate; E_s, E_{RC} are the modulus of elasticity of the steel substrate and RubCon layer, respectively; $k = 1 + (E_s/E_{RC})$, I_s, I_{RC} are the moments of inertia of the steel substrate and RubCon layer, respectively; F_{RC}, l_{RC} are the area and length of the RubCon layer, respectively; and t_s, t_{RC} are the thickness of the steel substrate and RubCon layer, respectively.

From the experimental results it may be deduced that at hardening of the RubCon polymer matrix, initial shrinkage stresses are $\sigma_{sh} \leq 1.6$ MA. It is worth noting that the shrinkage stress of other kinds of polymer concrete is much higher, e.g., 8–12 MPa for furfurol–acetone concrete, 10 MPa for epoxy concrete, and 5–6 MPa for polyester concrete. This means that the value of shrinkage stress of RubCon is insignificant in comparison with the compression strength of the material and for practical purposes does not influence structure behavior under loading. On the other hand, initial compressing stresses are an important concern in the combinations of RubCon and steel reinforcement. Shrinkage stresses compress a reinforcement bar, resulting in additional bond forces between reinforcement bar and matrix.

Mechanical Bond and Adhesion between Steel Reinforcement and the RubCon Matrix

Forces of mechanical bonds between the steel reinforcing bar and the RubCon matrix were experimentally determined with the help of the device shown in Figure 2.14.

A series of prismatic reinforced RubCon samples were produced. The reinforcement material was ribbed steel bar ø 8 mm. The height of samples or depth of a reinforcement anchorage was equaled from 35 up to 120 mm, that is, (4.3 ÷ 15)ø; the thickness of the protection layer was 8, 16, and 24 mm (1, 2, and 3) ø. Displacements of a reinforcing bar relatively the end face of the RubCon prism and load were measured during the tests. Results of these tests are shown in Figure 2.14 and Figure 2.15.

Destruction of samples with depth of anchorage of 35 and 40 mm (Figure 2.15) occurred as a result of shear of the reinforcement bar and the subsequent splitting of the RubCon prisms. At a depth of anchorage of 45 and 50 mm, destruction occurred due to rupture of the reinforcement. The initial displacements of a nonloaded end of the reinforcement bars were marked at stresses up to 50 MPa in all tested samples. Destruction of samples with a depth of anchorage of 60, 80, and 120 mm (Figure 2.14) occurred as a result of the rupture of the reinforcement bar. The initial displacements of a nonloaded end of the reinforcement bars were marked at stresses of 50–200 MPa. As this takes place, the final displacements of the bars did not depend on the thickness of the protection layer.

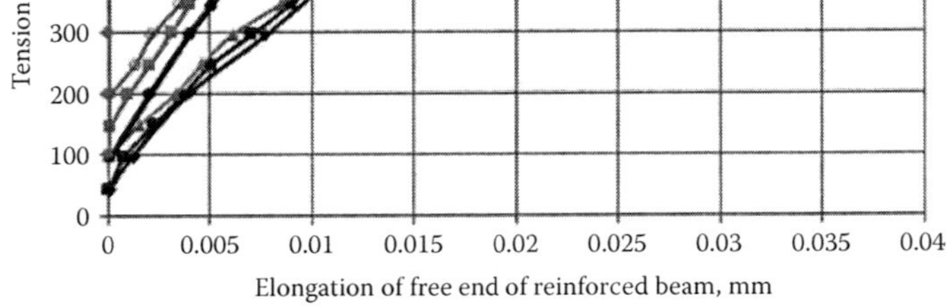

FIGURE 2.14 Draft of test gadget and displacements of non-loaded end of reinforcement bar in dependence of height of anchorage (a) and thickness of protection layer (b): (1, 2, 3) a = 60 mm, b = 8,16,24 mm correspondingly; (4, 5, 6) a = 80 mm, b = 8,16,24 mm correspondingly; (7) a = 120 mm, b = 8,16,24 mm correspondingly. (Reprinted from Yu. Potapov, O. Figovsky, Yu. Borisov, S. Pinaev, and D. Beilin, "Joint Work of Reinforcement and Polymer Concrete Matrix," *J. Scientific Israel Technological Advantages* 4, nos. 3–4 (2002): 14–19. With permission.)

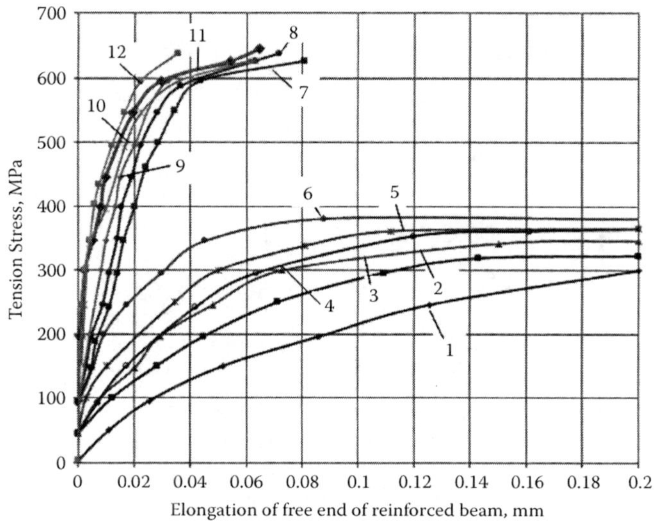

FIGURE 2.15 Displacements of nonloaded end of reinforcement bar in dependence of height of anchorage (a) and thickness of protection layer (b): (1, 2, 3) a = 35 mm, b = 8,16,24 mm correspondingly; (4, 5, 6) a = 40 mm, b = 8,16,24 mm correspondingly; (7, 8, 9) a = 45 mm, b = 8,16,24 mm correspondingly; (10, 11, 12) a = 50 mm, b = 8,16,24 mm correspondingly. (Reprinted from Yu. Potapov, O. Figovsky, Yu. Borisov, S. Pinaev, and D. Beilin, "Joint Work of Reinforcement and Polymer Concrete Matrix," *J. Scientific Israel Technological Advantages* 4, nos. 3–4 (2002): 14–19. With permission.)

We can conclude that the minimum thickness of the protection layer of reinforced RubCon should be not less than one diameter of the reinforcement bar.

The average bond stress between steel reinforcement and the RubCon matrix can be determined from the formula:

$$\tau_b = \sigma_r * A_r / u * l_r \tag{2.24}$$

where σ_r, A_r, l_r are the ultimate normal stress, cross-section area, and anchorage length (depth) of the reinforcement bar, respectively; and $u = \pi d$ is the perimeter of the reinforcement bar with diameter d.

Dependence of average bond stress τ_b on the anchorage depth of the reinforcement bar l_r is shown in Figure 2.16.

It is possible to see that at a depth of 45 mm,

$$\tau_b = \max \tau_b = 27.9 \div 28.6\text{MPa} \tag{2.24a}$$

One of the important factors in combinations of RubCon and steel reinforcement is an adhesive bond between these materials, which is caused by forces of intermolecular interaction. As research has shown [11], the adhesion of RubCon to steel is equal to 12.1 MPa, which means that approximately 0.8 is due to RubCon tensile strength.

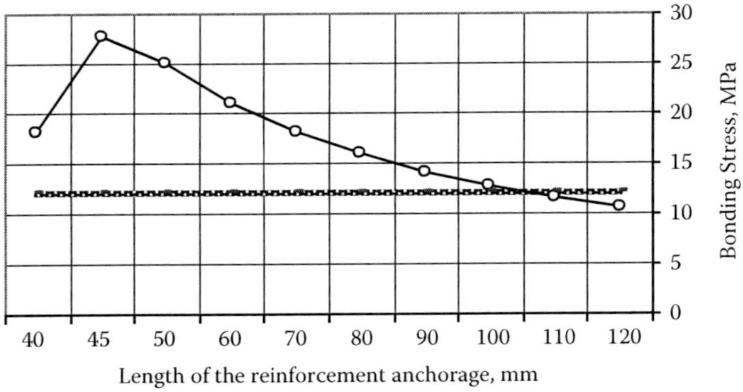

FIGURE 2.16 Dependence of average bond stress τ_b on anchorage depth of reinforcement bar l_r. (Reprinted from Yu. Potapov, O. Figovsky, Yu. Borisov, S. Pinaev, and D. Beilin, "Joint Work of Reinforcement and Polymer Concrete Matrix," *J. Scientific Israel Technological Advantages* 4, nos. 3–4 (2002): 14–19. With permission.)

It is necessary to note that this value exceeds the adhesion of furfurolacetate polymer concrete or usual cement concrete by approximately 10 times. Such high adhesion of the RubCon matrix to steel reinforcement is apparently related to the nature of the diene oligomer that is used as binder.

Load-Carrying Capacity at Eccentric Compression Load

Experimental research on the eccentrically compressed elements capacity was carried out on samples measuring 100 × 100 × 500 mm, reinforced by longitudinal rods of Ø8 mm and stirrups of Ø5 mm.

As a result of regression analysis of the experimental data, a mathematical model was obtained:

$$N_u = 3200e^{(0.23\mu - 2.15)} \tag{2.25}$$

where N_u is the limiting compressive load (kN), μ is the percentage of reinforcing (%), and e is the initial load eccentricity (cm).

From consideration of the response surface (Figure 2.17), it can be seen that the capacity of the investigated samples depends on eccentricity to a greater extent than on the reinforcement.

The analysis of Formula (2.25) shows that the increase in initial eccentricity of loading drastically reduces the loading capacity of an eccentrically compressed element of RubCon on the equation of power function. It is possible to note the "weak" dependence of the loading capacity on the reinforcing percent.

A design model for eccentric compressed elements from RubCon (Figure 2.18) was based on the following:

- Hypothesis of flat sections
- Physical nonlinearity of the material

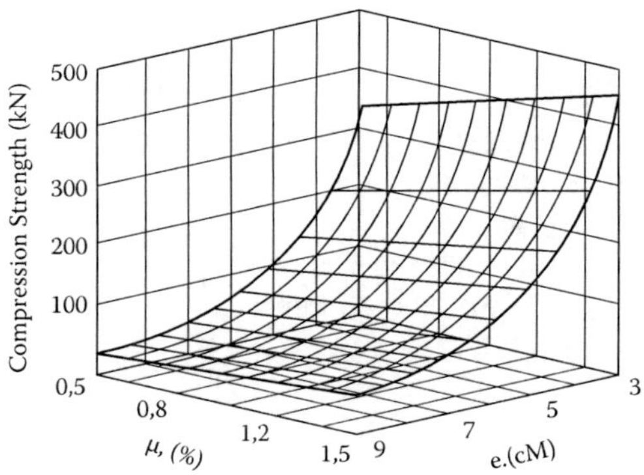

FIGURE 2.17 Relationship between compressive strength, percent of reinforcing and eccentricity. (Reprinted from Yu. Potapov, O. Figovsky, Yu. Borisov, S. Pinaev, and D. Beilin, "Stress–Strain State of Compressed Elements from Polymer Concrete," *J. Scientific Israel Technological Advantages* 4, nos. 3–4 (2002): 20–24. With permission.)

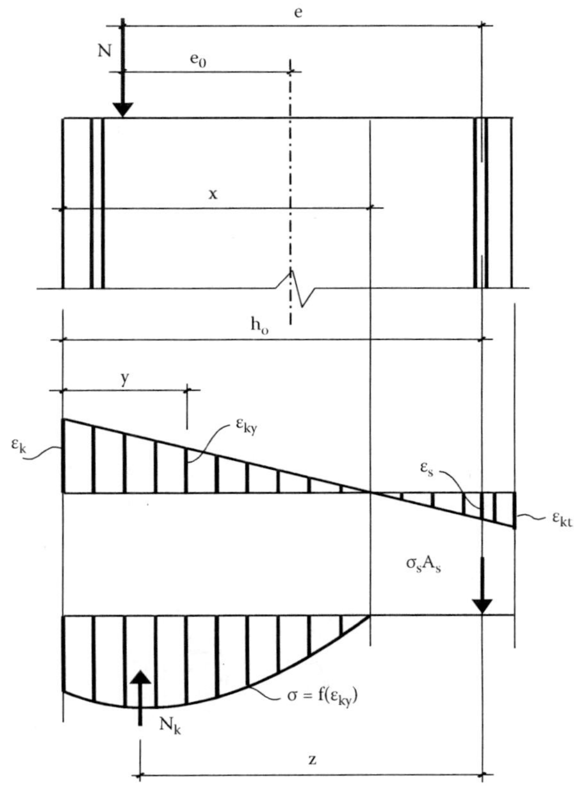

FIGURE 2.18 Model of stress–strain state of eccentrically compressed element. (Reprinted from Yu. Potapov, O. Figovsky, Yu. Borisov, S. Pinaev, and D. Beilin, "Stress–Strain State of Compressed Elements from Polymer Concrete," *J. Scientific Israel Technological Advantages* 4, nos. 3–4 (2002): 20–24. With permission.)

- Work of the tension zone in a condition approaching destruction is insignificant
- Position of the force system result is constant during all loading processes
- Exhaustion of loading capacity comes at achievement of a maximum on the diagram "bending moment—curvature"

Let's write down the equilibrium equations for a normal section of an eccentrically compressed RubCon short column (without buckling) at loading that approaches destruction:

$$\sum z = 0,\ N - N_k + \sigma_S A_S = 0; \tag{2.26}$$

$$\sum M_o = 0,\ Ne - N_k z_k = 0. \tag{2.27}$$

In view of nonlinear dependence between stress and strain, we have

$$N - b\int_0^x \sigma(y)dy + \sigma_S A_S = 0 \tag{2.28}$$

$$Ne - z_k b\int_0^x \sigma(y)dy = 0. \tag{2.29}$$

where y is the current coordinate of height of the cross section, and z is the shoulder of the internal force pair the forces, equal:

$$z_k = h_o - \frac{\int_0^x \sigma(y)\,y\,dy}{\int_0^x \sigma(y)dy} \tag{2.30}$$

Normal stress in a compressed zone may be computed from the uniaxial compression formula:

$$\sigma = \sigma_R\left[k\left(\frac{\varepsilon}{\varepsilon_R}\right) - \left(\frac{\varepsilon}{\varepsilon_R}\right)^2\right]. \tag{2.31}$$

with

$$\varepsilon_{ky} = \varepsilon_k \frac{x-y}{x}, \quad \varepsilon_s = \varepsilon_k \frac{h_o - x}{x}. \tag{2.32}$$

Substituting Equations (2.31) and (2.32) in the equilibrium Equations (2.28), (2.29) we have

$$N - bR_k \int_0^x \left[k\left(\frac{\varepsilon_k}{\varepsilon_R}\frac{x-y}{x}\right) - \left(\frac{\varepsilon_k}{\varepsilon_R}\frac{x-y}{x}\right)^2 \right] dy + E_s A_s \varepsilon_k \frac{h_o - x}{x} = 0; \quad (2.33)$$

$$Ne - bR_k \int_0^x \left[k\left(\frac{\varepsilon_k}{\varepsilon_R}\frac{x-y}{x}\right) - \left(\frac{\varepsilon_k}{\varepsilon_R}\frac{x-y}{x}\right)^2 \right] dy \times$$

$$\times \left(h_o - \frac{\int_0^x \left[k\left(\frac{\varepsilon_k}{\varepsilon_R}\frac{x-y}{x}\right) - \left(\frac{\varepsilon_k}{\varepsilon_R}\frac{x-y}{x}\right)^2 \right] y\,dy}{\int_0^x \left[k\left(\frac{\varepsilon_k}{\varepsilon_R}\frac{x-y}{x}\right) - \left(\frac{\varepsilon_k}{\varepsilon_R}\frac{x-y}{x}\right)^2 \right] dy} \right) = 0 \quad (2.34)$$

Supposing that loss of load capacity of an eccentrically compressed element is determined by the maximal bending moment, we shall add a third equation, describing the condition of a maximum:

$$\frac{dM}{d\varepsilon_k} = 0, \quad (2.35)$$

or

$$\frac{d}{d\varepsilon_k} \left\{ bR_k \int_0^x \left[k\left(\frac{\varepsilon_k}{\varepsilon_R}\frac{x-y}{x}\right) - \left(\frac{\varepsilon_k}{\varepsilon_R}\frac{x-y}{x}\right)^2 \right] dy \times \right.$$

$$\left. \times \left(h_o - \frac{\int_0^x \left[k\left(\frac{\varepsilon_k}{\varepsilon_R}\frac{x-y}{x}\right) - \left(\frac{\varepsilon_k}{\varepsilon_R}\frac{x-y}{x}\right)^2 \right] y\,dy}{\int_0^x \left[k\left(\frac{\varepsilon_k}{\varepsilon_R}\frac{x-y}{x}\right) - \left(\frac{\varepsilon_k}{\varepsilon_R}\frac{x-y}{x}\right)^2 \right] dy} \right) \right\} = 0. \quad (2.36)$$

After integration and transformations of Equations (2.31), (2.32), and (2.34), we shall write:

$$\begin{cases} N - \dfrac{bR_k}{3} \times \dfrac{x\left(3\varepsilon_R\varepsilon_k - \varepsilon_k^2\right)}{\varepsilon_R^2} + \sigma_s A_s = 0 \\ N \cdot e - \dfrac{bR_k}{12} \times \dfrac{12h_o\varepsilon_R\varepsilon_k x - 4h_o\varepsilon_k^2 x + \varepsilon_k^2 x^2 - 4\varepsilon_R\varepsilon_k x^2}{\varepsilon_R^2} = 0\,. \\ \varepsilon_k = \varepsilon_R \dfrac{6h_o - 2x}{4h_o - x} \end{cases} \quad (2.37)$$

We obtained the system of three equations with three unknowns, which can be solved for analytically, where N is the limiting compressive load; b is the width of the cross-section of the sample; R_k is the compressive strength of RubCon; x is the distance from the most compressed fiber to the neutral axis of the cross-section; ε_R is the deformation corresponding to maximal stress from the diagram $\sigma - \varepsilon$; ε_k is the deformation of extreme compressed fiber; σ_s is the tension stress; A_s is the tension zone area of the cross section; and h_o is the distance between centers of longitudinal reinforcements.

The results of experiment are in close agreement with the calculations of Formula (2.37).

Shear Stress and Crack Formation at Bend

Load-Carrying Capacity of RubCon Beams at Bend in the Transverse Forces Zone

Experimental research on the load-bearing capacity of RubCon beams at a bend was carried out on samples of beams measuring 6 × 12 × 140 cm. The draft of the samples and loading is shown in Figure 2.19.

Five series of samples were made, distinguished by their reinforcement. Beams of series A, B, and C had longitudinal ribbed reinforcement 2Ø12 (R_a = 6400 kg/cm^2); beams of series D and E had longitudinal ribbed reinforcement 2Ø12 (R_a = 3400 kg/cm^2).

Beams of series C and D were manufactured without transverse reinforcements. Samples of series A and E used wire stirrups Ø5 @ 5 cm (R_a = 3400 kg/cm^2). Beams of series B used wire stirrups Ø5 @ 3 cm (R_a = 3400 kg/cm^2). Beams of each sample series had action zones of transverse force of various lengths, namely, 12, 18, 24, 30, and 36 cm (a/h = 1, 1.5, 2, 2.5, and 3, respectively).

Beams were loaded using two symmetrically located concentrated loads. Such loading draft is appropriate for a so-called pure bend and is optimum for examining the influence of transverse forces in zones between supports and loads. The value of the transverse forces in these zones is constant, and bending moments are increased from zero on supports up to a maximum in cross-sections of a beam under loads. Beam samples were tested on a laboratory press. Loading was done with speeds of 500 kg/min up to

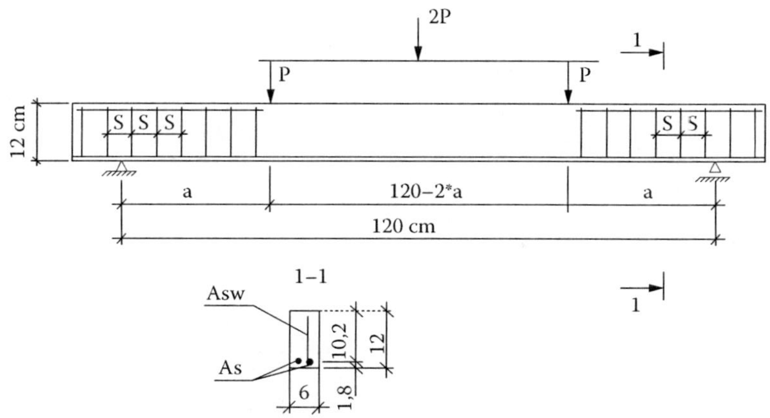

FIGURE 2.19 Test sample and draft of loading. (Reprinted from Yu. Potapov, O. Figovsky, Yu. Borisov, A. Polikushkin, and D. Beilin, "Influence of Shear Force on the Behavior of Polymer Concrete Beams at Bend," *J. Scientific Israel Technological Advantages* 4, nos. 3–4 (2002): 25–31. With permission.)

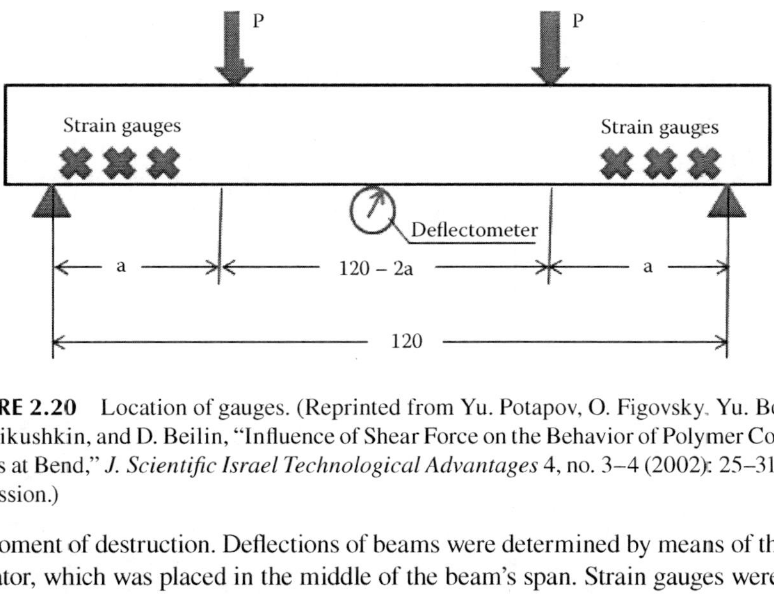

FIGURE 2.20 Location of gauges. (Reprinted from Yu. Potapov, O. Figovsky, Yu. Borisov, A. Polikushkin, and D. Beilin, "Influence of Shear Force on the Behavior of Polymer Concrete Beams at Bend," *J. Scientific Israel Technological Advantages* 4, no. 3–4 (2002): 25–31. With permission.)

the moment of destruction. Deflections of beams were determined by means of the dial indicator, which was placed in the middle of the beam's span. Strain gauges were used to measure RubCon deformations. Gauges were placed in a zone of transverse forces action along trajectories of the main compression and tensile stresses (Figure 2.20).

The analysis of the results shows the dependence of the load-bearing capacity of a beam on the length of a transverse force zone. For beams of all series, the dependencies are shown in Figures 2.21 and 2.22. It is possible to see that with increases of shear zone length and span of transverse reinforcement, the load-bearing capacity of bent beams decreases. This reduction becomes especially well marked for beams of series A–C (Figure 2.21) at shear zone length (1.5–1.75) h and for beams of a series D and E (Figure 2.22) at length $2h$.

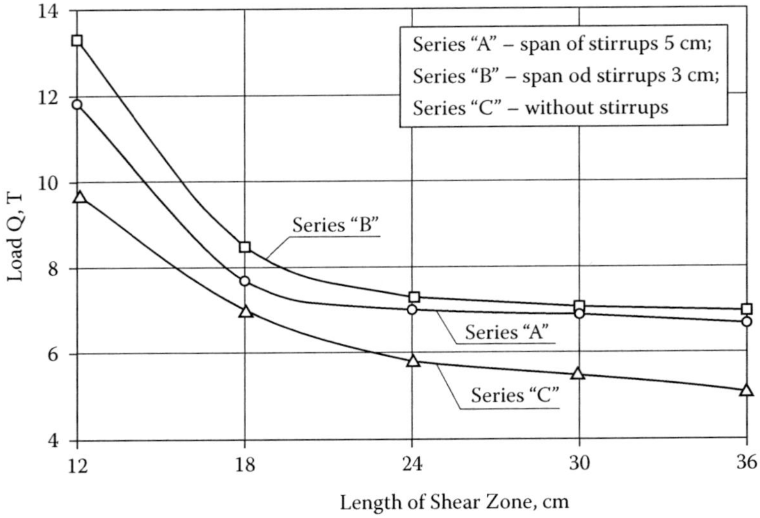

FIGURE 2.21 Relationship of the load bearing capacity of beams at bend and spans of stirrups and length of shear zone. Longitudinal reinforcement is 2Ø12 (R_a = 6400 kg/cm^2). (Reprinted from Yu. Potapov, O. Figovsky, Yu. Borisov, A. Polikushkin, and D. Beilin, "Influence of Shear Force on the Behavior of Polymer Concrete Beams at Bend," *J. Scientific Israel Technological Advantages* 4, nos. 3–4 (2002): 25–31. With permission.)

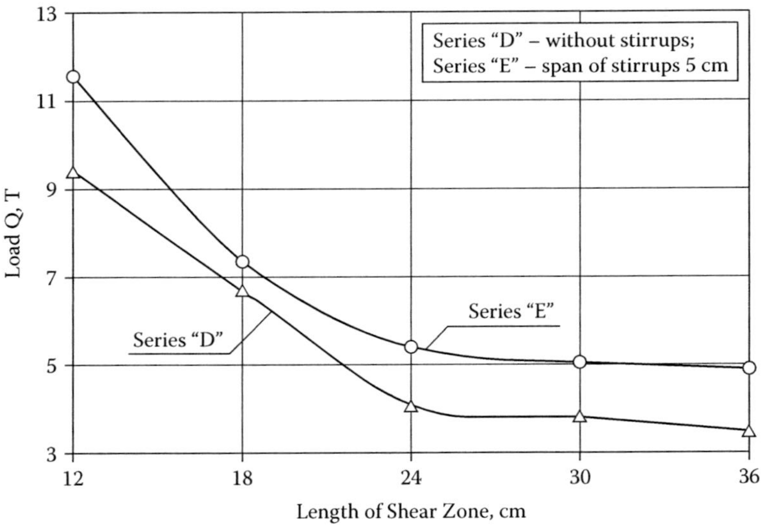

FIGURE 2.22 Relationship of the load bearing capacity of beams at bend and spans of stirrups and length of shear zone. Longitudinal reinforcement is 2Ø12 (R_a = 3400 kg/cm^2). (Reprinted from Yu. Potapov, O. Figovsky, Yu. Borisov, A. Polikushkin, and D. Beilin, "Influence of Shear Force on the Behavior of Polymer Concrete Beams at Bend," *J. Scientific Israel Technological Advantages* 4, nos. 3–4 (2002): 25–31. With permission.)

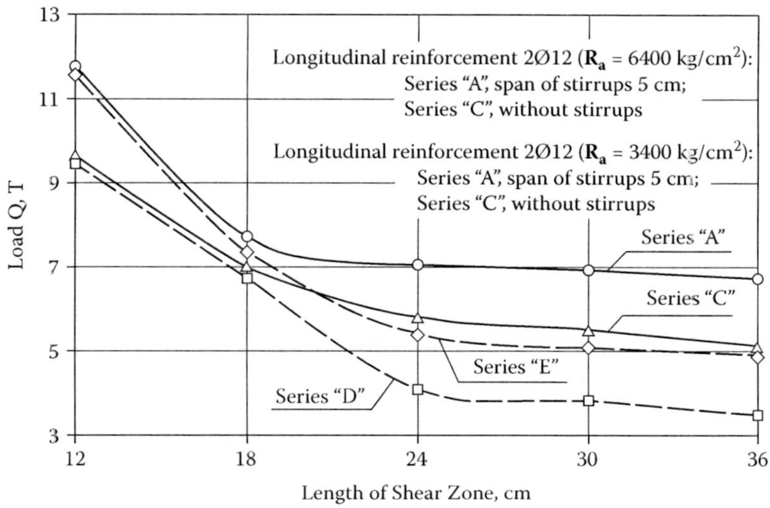

FIGURE 2.23 Influence of the strength of longitudinal reinforcement on load-bearing capacity of bent beams in shear zones. (Reprinted from Yu. Potapov, O. Figovsky, Yu. Borisov, A. Polikushkin, and D. Beilin, "Influence of Shear Force on the Behavior of Polymer Concrete Beams at Bend," *J. Scientific Israel Technological Advantages* 4, nos. 3–4 (2002): 25–31. With permission.)

To investigate the influence of longitudinal reinforcement strength on the load-bearing capacity of bent beams in shear zones, a comparative analysis of samples from series A and E on the one hand, and B and D on the other, was carried out. These pairs are distinguished only by design strength of a longitudinal reinforcement. Results of this research are shown in Figure 2.23.

Figure 2.23 shows that at rather short shear zone lengths ($a < 2h$), the influence of longitudinal reinforcement strength has practically no effect on the load-bearing capacity of beams (beams of series A–E and C–D). Longitudinal reinforcement in this zone perceives the transverse forces, but with increases of the shear zone, the longitudinal reinforcement in the greater degree works in a tension. It is obvious that the load-bearing capacity of beams under transverse forces with longitudinal reinforcement R_a = 6400 kg/cm^2 is higher than beams with reinforcement R_a = 3400 kg/cm^2.

Results of deflection measurements in the middle of a beam span (beam series A and E) are given in Figures 2.24 and 2.25. It is possible to see that series E beams with longitudinal reinforcement R_a = 3400 kg/cm^2 have the large deflections and smaller load-bearing capacity in comparison with series A beams with reinforcement R_a = 6400 kg/cm^2. It is necessary to note that these distinctions grow with an increase in the length of a shear zone due to the different deformation ability of the longitudinal reinforcements of these beams.

The increased deformation ability of the longitudinal reinforcement of beam series A contributes to plastic deformations and, consequently, to the large deflection values.

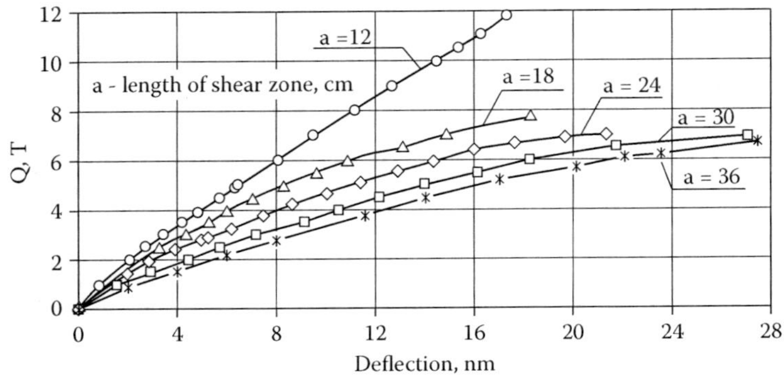

FIGURE 2.24 Deflections of beam series "A" with longitudinal reinforcements $R_a = 3400$ kg/cm^2 at different lengths of shear zone. (Reprinted from Yu. Potapov, O. Figovsky, Yu. Borisov, A. Polikushkin, and D. Beilin, "Influence of Shear Force on the Behavior of Polymer Concrete Beams at Bend," *J. Scientific Israel Technological Advantages* 4, nos. 3–4 (2002): 25–31. With permission.)

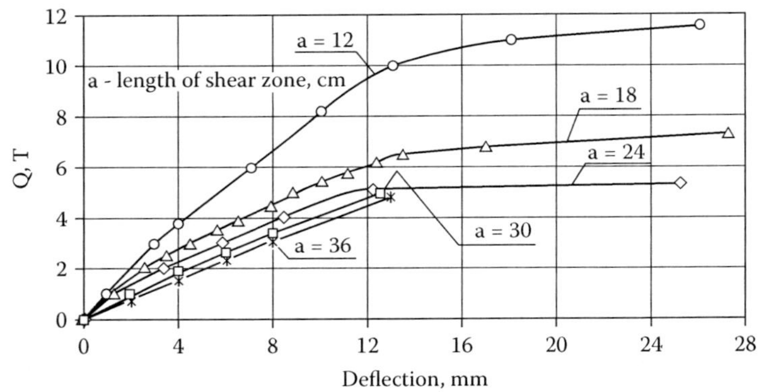

FIGURE 2.25 Deflections of beam series "E" with longitudinal reinforcements $R_a = 6400$ kg/cm^2 at different lengths of shear zone. (Reprinted from Yu. Potapov, O. Figovsky, Yu. Borisov, A. Polikushkin, and D. Beilin, "Influence of Shear Force on the Behavior of Polymer Concrete Beams at Bend," *J. Scientific Israel Technological Advantages* 4, nos. 3–4 (2002): 25–31. With permission.)

The received experimental data allow us to conclude that application of a longitudinal reinforcement $R_a = 6400$ kg/cm^2 has a favorable effect on the resistance of the shear beam zones at bend and promotes an increase of crack resistance and a reduction in the general deflection of a beam.

Phenomenological Model of Inclined Crack Formation and Destruction of RubCon

Let's consider the mechanism of inclined crack formation and destruction of RubCon material.

The inclined cracks that arise in shear zones under the main stress influence are distributed to approximately 2/3 of a beam height. In this case, the ends of a crack settle at some distance from the load and a support. Such cracks are characteristic for some beams with longitudinal reinforcement $R_a = 6400$ kg/cm^2. For the same beams, and for all beams with reinforcement $R_a = 3400$ kg/cm^2, crack formation occurs in most cases on the bottom side of the beam. An increase of loading on the crack is distributed along the main compression stress trajectory and its top is situated from a load center on approximately ¼ of the beam's height.

During the loading process, the appearance of horizontal cracks above reinforcement or along its axis was observed. These cracks were distributed from an inclined crack to a support. Formation of horizontal cracks preceded instant destruction of the beam.

Let's consider the forces exerted on the area of a beam near a support, limited to a line force action (Figure 2.26).

It is obvious that (Figure 2.26b):

$$Q * Q_k = N_s * N_k \tag{2.38}$$

where the first pair of forces (Figure 2.26c) influences the beam deflection, promoting an inclined crack formation. The force causing the separation of a concrete cover and formation of a horizontal crack (Figure 2.26d) is

$$N_s' = N_s * \sin \alpha. \tag{2.39}$$

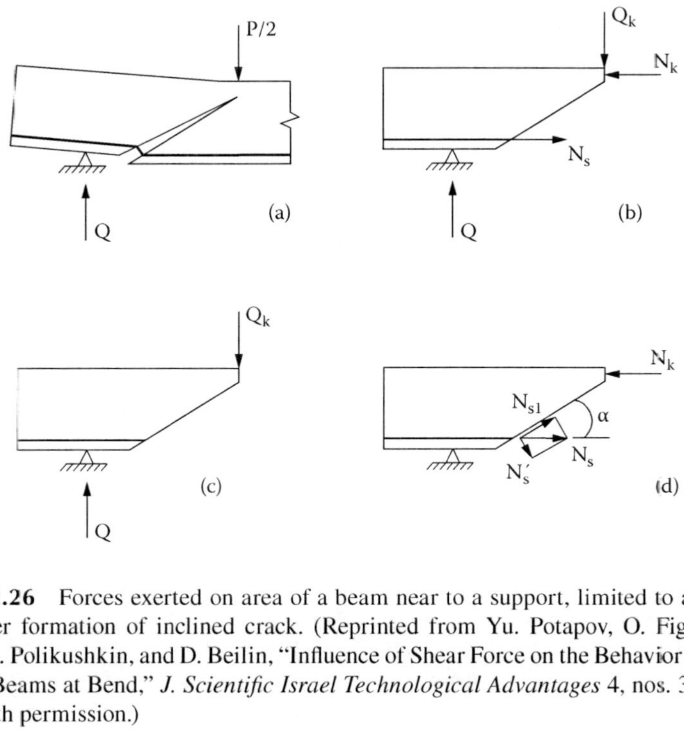

FIGURE 2.26 Forces exerted on area of a beam near to a support, limited to a line force action after formation of inclined crack. (Reprinted from Yu. Potapov, O. Figovsky, Yu. Borisov, A. Polikushkin, and D. Beilin, "Influence of Shear Force on the Behavior of Polymer Concrete Beams at Bend," *J. Scientific Israel Technological Advantages* 4, nos. 3–4 (2002): 25–31. With permission.)

The force N_s' causes a bend of the reinforcement bar and, in so doing, tensile stress can reach the yield point. When RubCon stress at the edge of the inclined crack adjoining the reinforcement reaches the ultimate tensile strength, there is a separation of the concrete cover and then there is destruction of the beam. The separation force is

$$N_s' = 0.5\, N_s * \sin 2\alpha. \tag{2.40}$$

It has been found experimentally that the crack inclination angle is between 20° and 45°, and hence the limits of separation force are from 0.32 N_s (for beams with large shear zone lengths) up to 0.5 N_s (for beams with small shear zone lengths).

Figure 2.27 shows the influence of stirrups on the load-bearing capacity of beams at bending for different shear zone lengths.

It can be seen that an increase in the transverse-reinforcing percentage results in some increase of load-bearing capacity of beams in a shear zone, especially at short shear zone lengths. The behavior of a beam without stirrups is similar to a truss with an intersecting lattice (Figure 2.28a) in which compression and tensile forces directed along trajectories of the main stresses (Figure 2.28b) are transferred on RubCon.

Occurrence of inclined cracks "cuts" the imaginary descending diagonals like a truss with ascending diagonals (Figure 2.28c). Stirrups stabilize the beam behavior even after appearance of an inclined crack. These stirrups fulfill the role of posts.

Fiber-Reinforced RubCon

Fiber composite materials have been widely used in construction in a variety of industries, such as aerospace, automotive, shipbuilding, and chemical processing, for

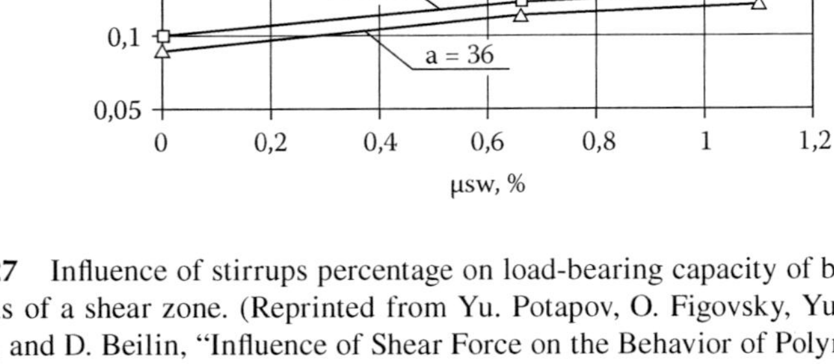

FIGURE 2.27 Influence of stirrups percentage on load-bearing capacity of beams for different lengths of a shear zone. (Reprinted from Yu. Potapov, O. Figovsky, Yu. Borisov, A. Polikushkin, and D. Beilin, "Influence of Shear Force on the Behavior of Polymer Concrete Beams at Bend," *J. Scientific Israel Technological Advantages* 4, nos. 3–4 (2002): 25–31. With permission.)

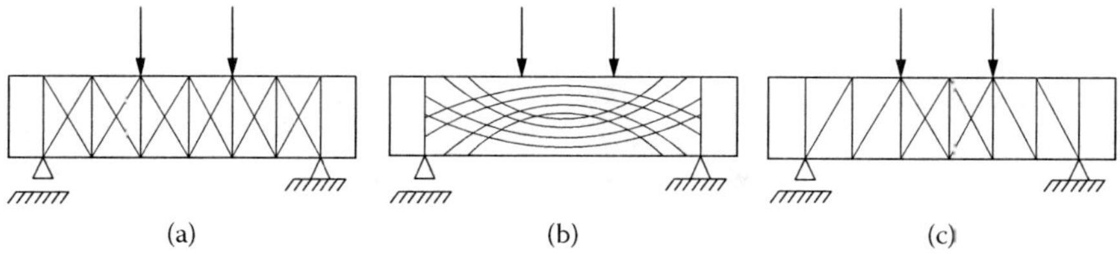

FIGURE 2.28 Diagrams of a RubCon beam behavior: (a) beam without cracks is similar to truss with intersecting lattice, (b) trajectories of the mean tensile and compressive stresses, (c) beam after cracks appearance is similar truss with ascending diagonals. (Reprinted from Yu. Potapov, O. Figovsky, Yu. Borisov, A. Polikushkin, and D. Beilin, "Influence of Shear Force on the Behavior of Polymer Concrete Beams at Bend," *J. Scientific Israel Technological Advantages* 4, nos. 3–4 (2002): 25–31. With permission.)

many years. Their application in civil engineering, however, has been limited, even though their high strength-to-weight ratio and excellent resistance to electrochemical corrosion make them attractive materials for structural applications. From this point of view, the new continuously reinforced polymeric concrete based on low molecular polybutadiene binder, with the best mechanical characteristics and chemical resistance in different aggressive environments, is rather perspective constructional material.

The maximum load-carrying capacity of fiber reinforcement is affected by fibers pulling out of the composite because fibers do not have a deformed surface like larger steel-reinforcing bars. This condition limits performance to a point that is far less than the yield strength of the fiber itself. This is important because some fibers are more "slippery" than others when used as reinforcement and will affect the toughness of the concrete product in which they are placed. In freshly mixed concrete, they increase resistance to plastic shrinkage. In hardened concrete, they improve strength and toughness, depending on the fiber type, shape, size, and amount.

Influence of Particle Size of Coarse Aggregates and Reinforcement Ratio of Fibrous RubCon Strength

Properties of a fibrous concrete as a composite material depend on the properties of its components, the major forms of which are steel or nonmetallic fibers. Because the Young's modulus of steel fibers exceeds the modulus of elasticity of concrete by 5 or 6 times, it is possible to increase the strength of concrete and its crack resistance. According to lab tests, small-diameter steel fibers are added to concrete to reduce its shrinkage cracking by more than 80%, and increase its toughness and long-term ability to withstand weather. The world experience of fibrous concrete application in construction shows the high technical and economic efficiency of this material.

Different factors, such as the reinforcement ratio, size, and form of fibers, and the particle size of a coarse aggregate, exert a great action on the mechanical properties of a continuously reinforced concrete. The influence of these parameters on strength and deformation characteristics of fibrous polymer concrete is investigated.

We examined RubCon that was hardened by continuously reinforced, arbitrarily located steel fibers. Fibers were made from chopped high-strength steel tire cord. They had a brass cover and wave shape form. Our preliminary research of the fibrous RubCon containing fine-grained gravel have shown that fibers with relative length L/d = 100 are located uniformly, without clots, and allow achievement of high strength and deformation characteristics.

A full two-factor experiment was carried out. The varied parameters were the reinforcement ratio $\boldsymbol{\mu}$ and the particle size of coarse aggregate $\boldsymbol{a}$; response functions were the tensile, compressive, and bending strength of the polymer concrete samples measuring 4 × 4 × 16 cm and 4 × 4 × 40 cm. During the experiment, the reinforcement ratio and particle size were changed from 0% (without reinforcement) to 3% in 1% increments, and from 0 (without crushed stone) to 15-mm in 5-mm steps, respectively.

Results of the experiments are illustrated in Figures 2.29 through 2.31.

It can be seen that, with increased particle size of coarse aggregate at a constant reinforcement ratio, the RubCon strength characteristics are decreased. This tendency is especially clearly shown for tensile strength. The increase of strength characteristics with growth of reinforcing ratio in all range values was expected, given the particle size of coarse aggregate.

The experiments have made it possible to propose a practical combination of reinforcing ratio and coarse aggregate quality assignment depending on the required

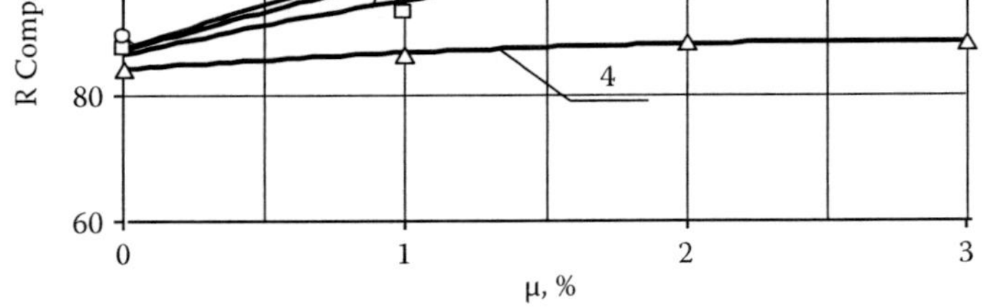

FIGURE 2.29 Influence reinforcement ratio and particle size of coarse aggregate *on compression strength* of fibrous RubCon: (1) without crushed stone, (2) with crushed stone up to 5 mm. (3) with crushed stone up to 10 mm, (4) with crushed stone up to 15 mm. (Reprinted from Yu. Potapov, Yu. Borisov, D. Panfilov, O. Figovsky, and D. Beilin, "Research of Polymer Concrete Based on Low Molecular Polybutadiene, Part VI: Influence of Particle Size of Coarse Aggregates and Reinforcement Ratio on the Fibrous Polymer Concrete Strength," *J. Scientific Israel Technological Advantages* 6, nos. 3–4 (2004): 67–70. With permission.)

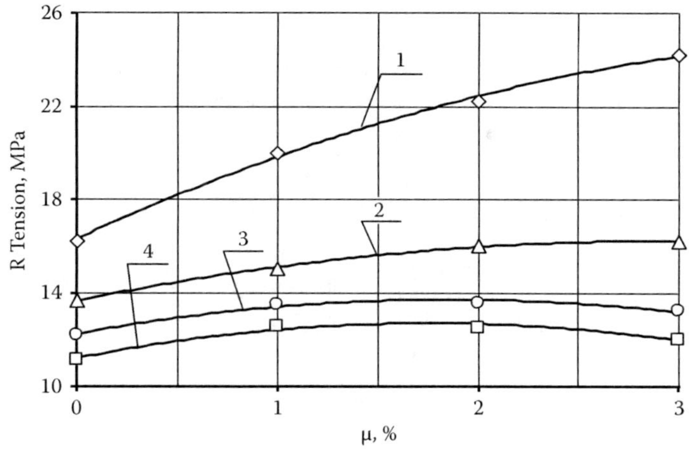

FIGURE 2.30 Influence reinforcement ratio and particle size of coarse aggregate *on tensile strength* of fibrous RubCon: (1) without crushed stone, (2) with crushed stone up to 5 mm, (3) with crushed stone up to 10 mm, (4) with crushed stone up to 15 mm. (Reprinted from Yu. Potapov, Yu. Borisov, D. Panfilov, O. Figovsky, and D. Beilin, "Research of Polymer Concrete Based on Low Molecular Polybutadiene, Part VI: Influence of Particle Size of Coarse Aggregates and Reinforcement Ratio on the Fibrous Polymer Concrete Strength," *J. Scientific Israel Technological Advantages* 6, nos. 3–4 (2004): 67–70. With permission.)

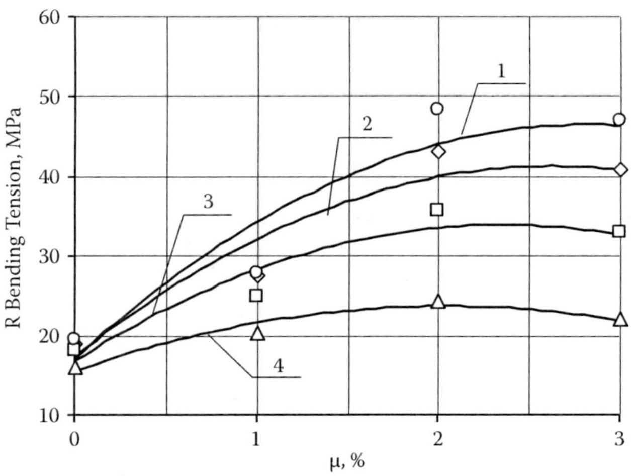

FIGURE 2.31 Influence reinforcement ratio and particle size of coarse aggregate *on bending strength* of fibrous RubCon: (1) without crushed stone, (2) with crushed stone up to 5 mm. (3) with crushed stone up to 10 mm, (4) with crushed stone up to 15 mm. (Reprinted from Yu. Potapov, Yu. Borisov, D. Panfilov, O. Figovsky, and D. Beilin, "Research of Polymer Concrete Based on Low Molecular Polybutadiene, Part VI: Influence of Particle Size of Coarse Aggregates and Reinforcement Ratio on the Fibrous Polymer Concrete Strength," *J. Scientific Israel Technological Advantages* 6, nos. 3–4 (2004): 67–70. With permission.)

strength characteristics of RubCon. The compression, tensile, and bending strength dependences of *a* and μ values are shown in Figures 2.32 through 2.34.

It is worth noting that at $\mu > 2\%$, the strength growth is slowed down because of clotting of fibers in the process of the RubCon mixing preparation. Clotting makes the production process more complicated and creates many defects in the RubCon structure.

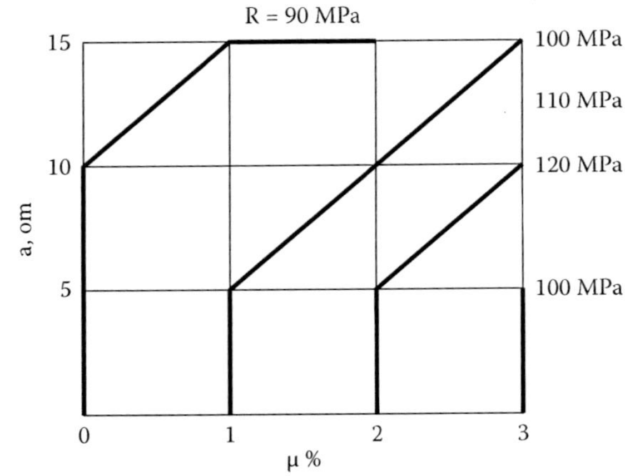

FIGURE 2.32 The RubCon *compression strength* (MPa) dependence of reinforcement ratio (%) and particle size of coarse aggregate (cm). (Reprinted from Yu. Potapov, Yu. Borisov, D. Panfilov, O. Figovsky, and D. Beilin, "Research of Polymer Concrete Based on Low Molecular Polybutadiene, Part VI: Influence of Particle Size of Coarse Aggregates and Reinforcement Ratio on the Fibrous Polymer Concrete Strength," *J. Scientific Israel Technological Advantages* 6, nos. 3–4 (2004): 67–70. With permission.)

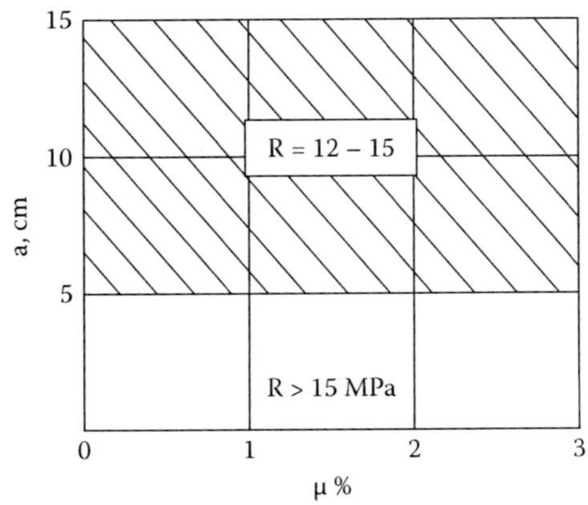

FIGURE 2.33 The RubCon *tensile strength* (MPa) dependence of reinforcement ratio (%) and particle size of coarse aggregate (cm). (Reprinted from Yu. Potapov, Yu. Borisov, D. Panfilov, O. Figovsky, and D. Beilin, "Research of Polymer Concrete Based on Low Molecular Polybutadiene, Part VI: Influence of Particle Size of Coarse Aggregates and Reinforcement Ratio on the Fibrous Polymer Concrete Strength," *J. Scientific Israel Technological Advantages* 6, nos. 3–4 (2004): 67–70. With permission.)

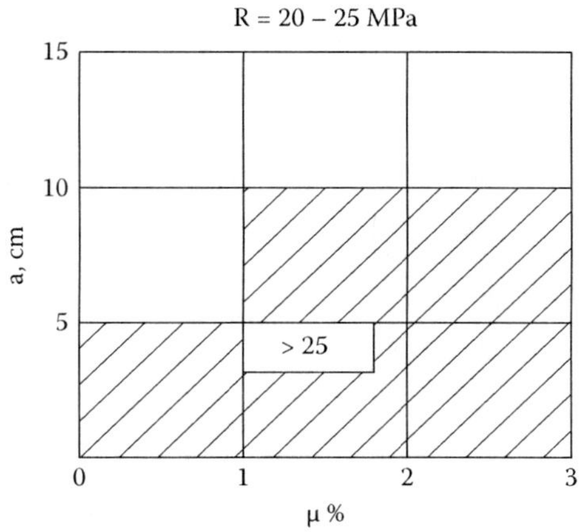

FIGURE 2.34 The RubCon *bending strength (MPa)* dependence of reinforcement ratio (%) and particle size of coarse aggregate (cm). (Reprinted from Yu. Potapov, Yu. Borisov, D. Panfilov, O. Figovsky, and D. Beilin, "Research of Polymer Concrete Based on Low Molecular Polybutadiene, Part VI: Influence of Particle Size of Coarse Aggregates and Reinforcement Ratio on the Fibrous Polymer Concrete Strength," *J. Scientific Israel Technological Advantages* 6, nos. 3–4 (2004): 67–70. With permission.)

Strength of Fibrous RubCon with Various Kinds of Fibers

Materials used for fiber reinforcement include acrylic, asbestos, cotton, glass, nylon, polyester, polyethylene, polypropylene, rayon, rock wool, and steel. Of these, acid-resistive glass and steel fibers have received the most attention. Plastic fibers have shown to be of little value in reinforcing concrete until only recently. Natural fibers are subject to alkali attack and are also determined to have little value. Nylon is currently making an appearance in slab-on-grade technology. Most of the test data, however, centers around the use of steel fibers and glass fibers. The influence of various kinds of fibers, reinforcement ratios and fiber aspect ratio on compressive, tensile, and bending strength of RubCon is examined. We used steel, polypropylene, rough basalt, and glass fibers in different amounts and sizes. The steel fibers with brass cover were manufactured from high-strength metal tire cord. Fibers are made by the chopping of cord and the subsequent defibering of the separate fibers in a forced mixer. Such fibers have a wave shape.

Experimental investigations were performed with fiber-reinforced RubCon samples measuring 4 × 4 × 16 cm. Various fiber materials were used: steel, polypropylene, rough basalt, and glass. The reinforcement ratio for appropriate fiber materials was adopted according to publicized research [4,11]. The optimal volume of fiber content is 1.0%–1.5% for steel fibers and 1.0–5.0% for glass fibers.

To determine the influence of reinforcement on the strength characteristics of RubCon, control samples without reinforcement (size 4 × 4 × 16 cm) were prepared. Figure 2.35 illustrates the dependence tension at bending of RubCon samples reinforced by different kinds of fibers. One can see that the bending

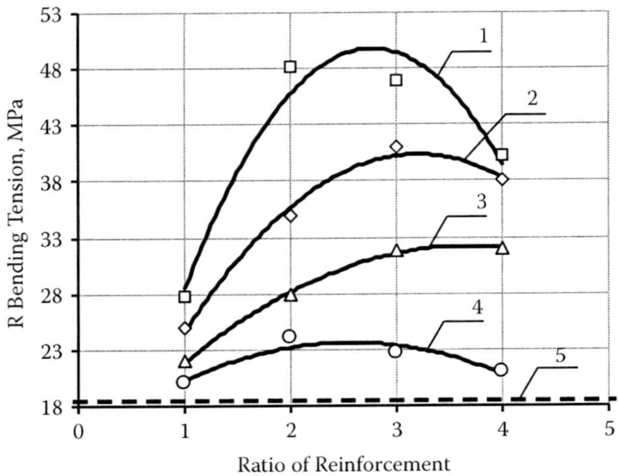

FIGURE 2.35 *RubCon*-bending tension dependence of the reinforcement ratio and the kind of fibers: (1) steel fibers, (2) polypropylene fibers, (3) rough basalts fibers, (4) glass fibers, (5) control sample without reinforcement. (Reprinted from Yu. Potapov, Yu. Borisov, D. Panfilov, O. Figovsky, and D. Beilin, "Research of Polymer Concrete Based on Low Molecular Polybutadiene, Part VII: Strength of Continuously Reinforced Polymer Concrete with Various Kinds of Fibers," *J. Scientific Israel Technological Advantages* 6, nos. 3–4 (2004): 71–74. With permission.)

strength of fibrous reinforcement samples strikingly differs from strength samples without reinforcement. In this case, steel fibers from metal cord provide the highest bending tension. It is worth noting that at an increased reinforcement ratio, the bending strength of RubCon decreases due to fiber clotting; our earlier experiments also confirm this.

In the course of our tests we observed various types of RubCon sample destruction depending on the fiber reinforcement material. The application of rough basalt or glass fibers results in brittle destruction of samples, which occurs after fiber breakage. The crack formation moment coincides with the breaking of fibers. It should be noted that crack formation of RubCon samples reinforced with steel or polypropylene fibers occurs much earlier than the fracture of fibers.

One of the major areas of RubCon application is in structures that operate in aggressive environments where crack resistance of the material is important. Our experiments have shown that steel fibers show the greatest opportunity to increase crack resistance and allow production of a material with high elastic-plastic properties. For this reason, we have undertaken research on the influence of fiber reinforcement ratio and aspect ratio on RubCon strength at compression, tension, and bend.

A full two-factor experiment was carried out. The varied parameters were the reinforcement ratio μ and aspect ratio (relative length) of the fibers ***L/d***; the response functions were tensile, compressive, and bending strength of the RubCon samples. During the experiment, the reinforcement ratio and aspect ratio were changed from 1% to 4% in 1% increments, and from 40 to 130 in increments of 30, respectively.

The results of the experiments are illustrated in Figures 2.36–2.38.

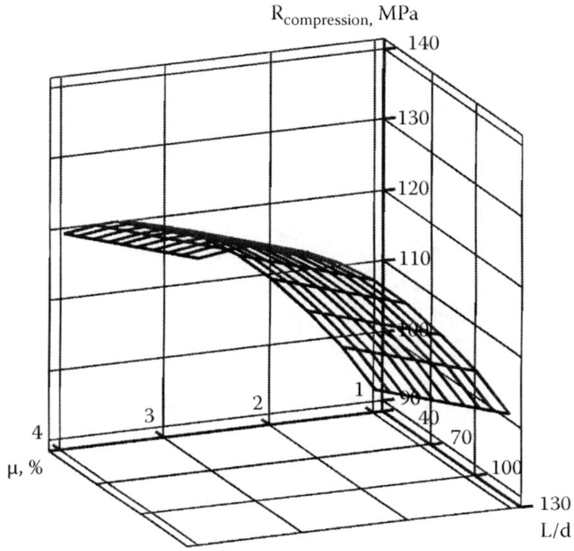

FIGURE 2.36 Influence of reinforcement ratio μ and aspect ratio *L/d* of the steel fibers *on compression strength* of fibrous RubCon. (Reprinted from Yu. Potapov, Yu. Borisov, D. Panfilov, O. Figovsky, and D. Beilin, "Research of Polymer Concrete Based on Low Molecular Polybutadiene, Part VII: Strength of Continuously Reinforced Polymer Concrete with Various Kinds of Fibers," *J. Scientific Israel Technological Advantages* 6, nos. 3–4 (2004): 71–74. With permission.)

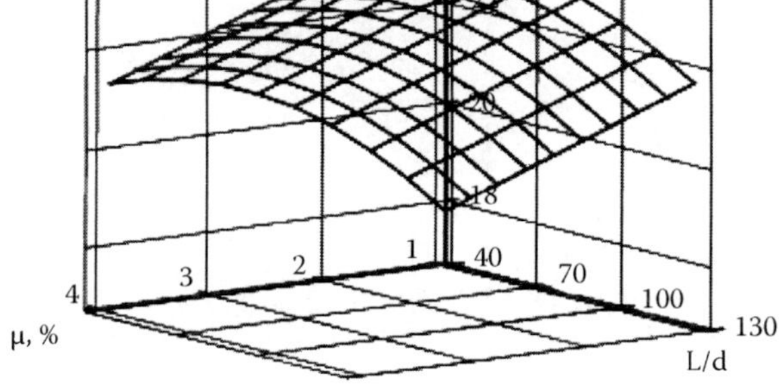

FIGURE 2.37 Influence of reinforcement ratio μ and aspect ratio *L/d* of the steel fibers *on tensile strength* of fibrous RubCon. (Reprinted from Yu. Potapov, Yu. Borisov, D. Panfilov, O. Figovsky, and D. Beilin, "Research of Polymer Concrete Based on Low Molecular Polybutadiene, Part VII: Strength of Continuously Reinforced Polymer Concrete with Various Kinds of Fibers," *J. Scientific Israel Technological Advantages* 6, nos. 3–4 (2004): 71–74. With permission.)

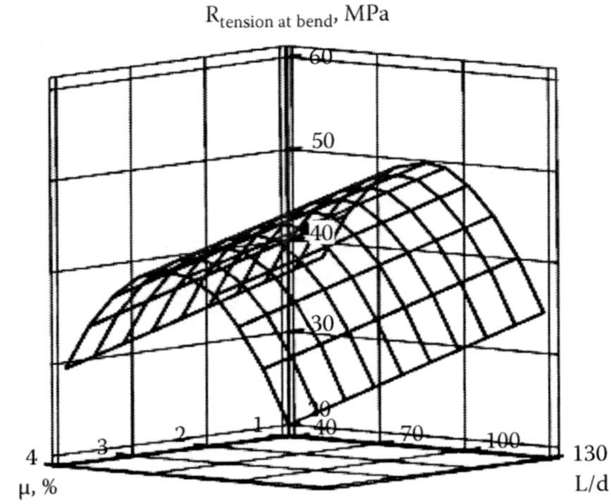

FIGURE 2.38 Influence of reinforcement ratio μ and aspect ratio *L/d* of the steel fibers *on bending strength* of fibrous RubCon. (Reprinted from Yu. Potapov, Yu. Borisov, D. Panfilov, O. Figovsky, and D. Beilin, "Research of Polymer Concrete Based on Low Molecular Polybutadiene, Part VII: Strength of Continuously Reinforced Polymer Concrete with Various Kinds of Fibers," *J. Scientific Israel Technological Advantages* 6, nos. 3–4 (2004): 71–74. With permission.)

The analysis of the results shows that with increased relative length of the fiber ***L/d***, the strength of RubCon increases, reaches a maximum, and is reduced. It is especially appreciable in the case of bending strength (Figure 2.38). The explanation of this phenomenon is connected to the interaction of a plastic matrix and rigid elastic fibers. It should be borne in mind that the modulus of elasticity of the steel fibers surpasses the modulus of elasticity of RubCon by 5 or 6 times. During the loading process at a large enough relation *L/d*, fibers "slip" from the polybutadiene binder due to a decrease of adhesion between the RubCon matrix and fibers.

The strength characteristics of steel fiber–reinforced RubCon depend on the reinforcement ratio μ. The results of experiments show that an optimum saturation by fiber reinforcement is in the range of 2%–3%. Fiber clotting is observed at greater volumes of reinforcement, and results in a decrease in *RubCon* strength and serious constructive defects.

During experiments, we investigated the formation and development of cracks in RubCon samples. Crack formation took place at *L/d* = 40, 70, 100, 130 and μ = 1% and at *L/d* = 40 and μ = 1, 2, 3%. In that case, $N_{crc}/N_u \approx 1$ where N_{crc} is the load crack formation, N_u is the ultimate load. Crack formation load is decreased at large values of *L/d* and μ ($N_{crc}/N_u \approx 0.66$).

Summary

- RubCon structures reinforced with steel-ribbed bars have a high bond stress due to adhesion and forces of mechanical bond.

- The behavior of bent RubCon structural elements is essentially similar to that of a reinforced concrete beam, which means that RubCon elements can be designed in the same way as reinforced concrete beams.
- Fiber production from tire industry waste for the manufacture of composite materials with continuous reinforcement is possible and expedient.
- The increase of the particle size of coarse aggregate at a constant reinforcement ratio results in a decrease of RubCon compression, tensile, and bending strength.
- The optimal composition of continuously reinforced RubCon from a strength point of view is: reinforcement ratio $\mu = 2\%$; maximal particle size of coarse aggregates $a = 5$ mm. At $\mu > 2\%$, clotting of fibers is observed, which results in some structural defects.
- Optimal parameters of steel fiber reinforcement are: reinforcement ratio $\mu = 2\%$; relative length of fiber $L/d = 100$.
- Application of steel fiber with the relative length ***L/d* > *100*** is not efficient.

RUBCON CREEP [1,3,18–20]

RubCon Sample Materials

Polymer concrete based on two kinds of liquid rubbers was investigated: type A, low molecular polybutadiene (Butarez®, Liten® [1,4-cis 25%–39%; 1,4-trans 35%–40%, 1,2-vinyl 28%–35%]) and type B, stereoregular low molecular rubber (Polyoil 110/130®, Ricon® (1,4-cis 70%–80%; 1,4-trans 20%–30%, 1,2-vinyl 1%–2%). The main physical–mechanical properties are shown in Table 2.5.

Creep Deformation of Plain RubCon at Long-Term Compressive Load

RubCon application in load-bearing structures is directly connected to the stress–strain state of a material under long-term loading action (creep). Creep of RubCon at compression was determined on samples measuring 40 × 40 × 160 mm.

Two series of RubCon samples of type A and type B were tested. The creep curves are shown in Figures 2.39 and 2.40.

The analysis of the resulting curves (Figure 2.39) shows that deformations of creep at a stress of ~75% from compression strength of samples ($\sigma_c = 110$ MPa) are damped out in a range from 20 days (sample no. 11, 20% of σ_c) up to 280 days (sample no. 4, 75% of σ_c). At higher stress values, continuous growth of creep deformations up to destruction (samples no. 1, 2, 3) is marked. Characteristic curves of creep deformation of type B samples ($\sigma_c = 58.4$MPa) are illustrated in Figure 2.40. One can see that the deformation–time ratio of these samples are similar to the samples of type A.

Dependencies of deformation on stress are given in Figure 2.41 for two types of RubCon samples at short-time and long-term loading. It is obvious that at long-term loading, these dependences have a nonlinear character.

Based on these experiments, the stress–strain relationship can be written as follows:

$$\sigma_0 = \varepsilon E - (\varepsilon E)^2/2\sigma \tag{2.41}$$

TABLE 2.5
Main Properties of the RubCon Samples

Properties		RubCon Type A	RubCon TYPE B
Average density (kg/m^3)		2050 … 2350	
Strength (MPa) at	Compression	80 …110	55 … 90
	Tension	10 … 20	9 …18
	Bend	20 … 33	18 … 28
Module of elasticity (MPa$_*$10^4)		1.2 … 3.0	1.2 … 2.5
Poisson's ratio		0.26	0.27
Coefficient of linear extension (10^{-5}$_*$°C^{-1})		1.2–1.6	
Water absorption (%)		0.05	
Adhesion to steel (Mpa)		15–25	
Coefficient of chemical resistance	30% Sulfuric acid	0.95	—
	20% Sulfuric acid	—	0.98
	10% Lactic acid	0.95	0.965
	20% Potassium hydroxide	—	0.98
	10% Sodium hydroxide	0.97	—
	Water	0.995	0.99

Source: Reprinted from Yu. Potapov, O. Figovsky, Yu. Borisov, S. Pinaev, and D. Beilin, "Creep of Polymer Concrete at Compressive Loading," *J. Scientific Israel Technological Advantages* 5, nos. 1–2 (2003): 1–10. With permission.

where σ_0 is the stress at long-term loading; ε is the strain at long-term loading; σ' is the coefficient of nonlinearity; and E is the modulus of elasticity.

As in other polymers, rubber concretes form three kinds of deformation: elastic, viscous, and highly elastic in a stress field (Figure 2.42).

According to Maxwell's rheological model of long-term action of constant loading, the stress σ_0 and viscous deformations ε_v are characterized by the equations:

$$\sigma_0 = \eta_1\, d\varepsilon_v/dt;\ \varepsilon_v = \sigma_0 *t/\eta_1 \tag{2.42}$$

In the case of highly elastic deformation, Binham's equation is useful:

$$\varepsilon_e = \sigma_0/E_1\,(1 - \exp\,(-E_1\, t\,/\eta) \tag{2.43}$$

As is known, a highly elastic deformation is accompanied by redistribution of internal stresses and is always associated with viscous deformation. Because of this, for full deformation of a viscous polymer we will have a rheological model (Figure 2.42c), which is governed by the equation:

$$\varepsilon_{ve} = \sigma_0 *t/\eta_1 + \sigma_0/E_1\,(1 - \exp\,(-E_1\, t\,/\eta). \tag{2.44}$$

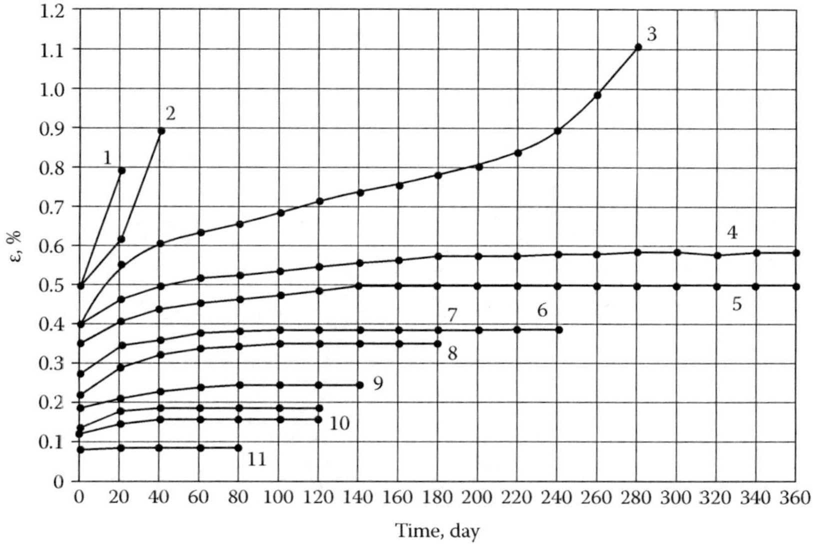

FIGURE 2.39 Creep curves of *RubCon* type A samples: (1) $\sigma_0 = 94.5$ MPa; $\varepsilon_0 = 0.428\%$; (2) $\sigma_0 = 89.3$ MPa; $\varepsilon_0 = 0.394\%$; (3) $\sigma_0 = 84$ MPa; $\varepsilon_0 = 0.344\%$; (4) $\sigma_0 = 78$ MPa; $\varepsilon_0 = 0.321\%$; (5) $\sigma_0 = 73.5$ MPa; $\varepsilon_0 = 0.288\%$; (6) $\sigma_0 = 68.3$ MPa; $\varepsilon_0 = 0.231\%$; (7) $\sigma_0 = 63$ MPa; $\varepsilon_0 = 0.203\%$; (8) $\sigma_0 = 52.5$ MPa; $\varepsilon_0 = 0.195\%$; (9) $\sigma_0 = 42$ MPa; $\varepsilon_0 = 0.148\%$; (10) $\sigma_0 = 31.5$ MPa; $\varepsilon_0 =.132\%$; (11) $\sigma_0 = 21$ MPa; $\varepsilon_0 = 0.076\%$. (Reprinted from Yu. Potapov, O. Figovsky, Yu. Borisov, S. Pinaev, and D. Beilin, "Creep of Polymer Concrete at Compressive Loading," *J. Scientific Israel Technological Advantages* 5, nos. 1–2 (2003): 1–10. With permission.)

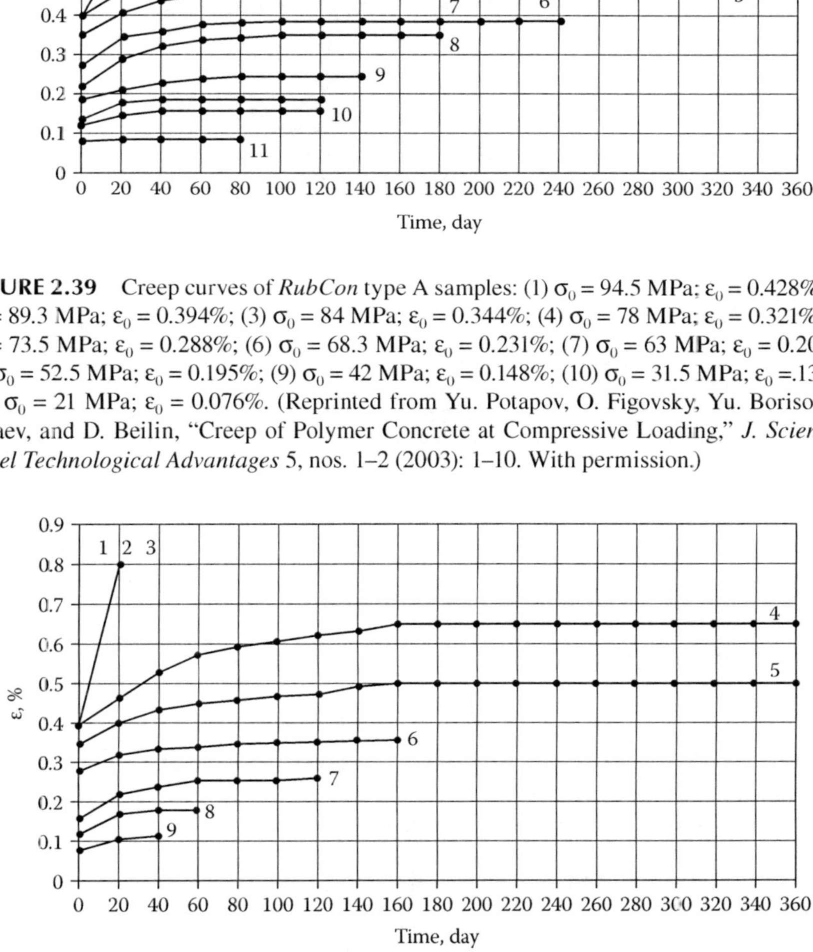

FIGURE 2.40 Creep curves of *RubCon* type B samples: (1) $\sigma_0 = 52.6$ MPa; $\varepsilon_0 = 0.35\%$; (2) $\sigma_0 = 49.7$ MPa; $\varepsilon_0 = 0.3\%$; (3) $\sigma_0 = 46.8$ MPa; $\varepsilon_0 = 0.28\%$; (4) $\sigma_0 = 43.8$ MPa; $\varepsilon_0 = 0.26\%$; (5) $\sigma_0 = 40.9$ MPa; $\varepsilon_0 = 0.225\%$; (6) $\sigma_0 = 38$ MPa; $\varepsilon_0 = 0.22\%$; (7) $\sigma_0 = 35.1$ MPa; $\varepsilon_0 = 0.18\%$; (8) $\sigma_0 = 29.2$ MPa; $\varepsilon_0 = 0.16\%$; (9) $\sigma_0 = 23.4$ MPa; $\varepsilon_0 = 0.09\%$. (Reprinted from Yu. Potapov, O. Figovsky, Yu. Borisov, S. Pinaev, and D. Beilin, "Creep of Polymer Concrete at Compressive Loading," *J. Scientific Israel Technological Advantages* 5, nos. 1–2 (2003): 1–10. With permission.)

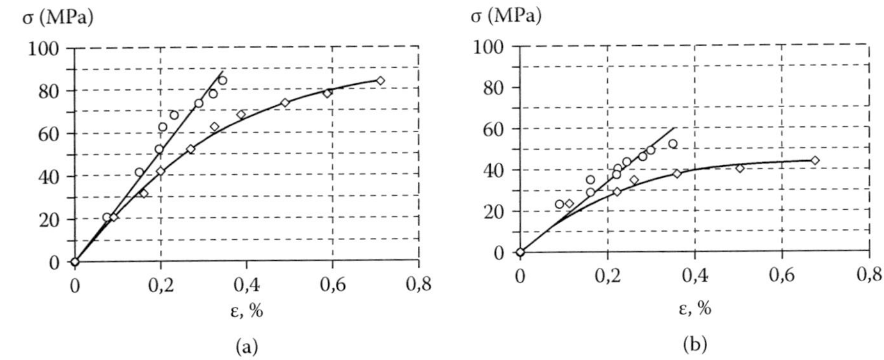

FIGURE 2.41 Stress–strain relationship at short time (continuous line) and long-time (dashed line) loading. (a) For *RubCon* of type A, (b) for *RubCon* of type B. (Reprinted from Yu. Potapov, O. Figovsky, Yu. Borisov, S. Pinaev, and D. Beilin, "Creep of Polymer Concrete at Compressive Loading," *J. Scientific Israel Technological Advantages* 5, nos. 1–2 (2003): 1–10. With permission.)

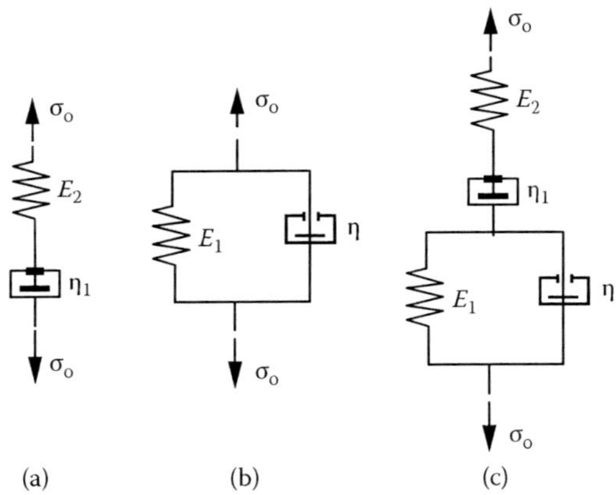

FIGURE 2.42 Rheological models: (a) viscous body, (b) elastic-viscous body, (c) viscous-elastic body. (Reprinted from Yu. Potapov, O. Figovsky, Yu. Borisov, S. Pinaev, and D. Beilin, "Creep of Polymer Concrete at Compressive Loading," *J. Scientific Israel Technological Advantages* 5, nos. 1–2 (2003): 1–10. With permission.)

As rubber concrete has long-term strength that is distinct from zero and creep damped out at compression, it is necessary to enter element E representing an elastic molecular skeleton of a material into the rheological material circuit (Figure 2.43).

However, if deformations of nonlinear creep depend on stress, it is impossible to analyze the creep process of a material on a mechanical model base. As experiments have shown, creep deformations of both RubCon sample types align with the equation of a square parabola, which results from the structural diagram (Figure 2.44). This structural diagram looks like an inclined straight line, passing through the beginning of coordinates.

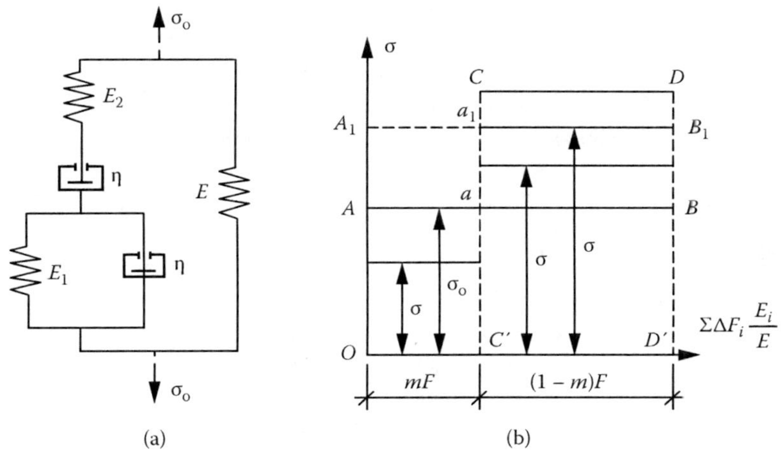

FIGURE 2.43 Modified rheological model (a) and structural diagram (b) of a viscous-elastic body; *mF, (mF* – 1*)* are the cross section of viscous and elastic phase particles, respectively. (Reprinted from Yu. Potapov, O. Figovsky, Yu. Borisov, S. Pinaev, and D. Beilin, "Creep of Polymer Concrete at Compressive Loading," *J. Scientific Israel Technological Advantages* 5, nos. 1–2 (2003): 1–10. With permission.)

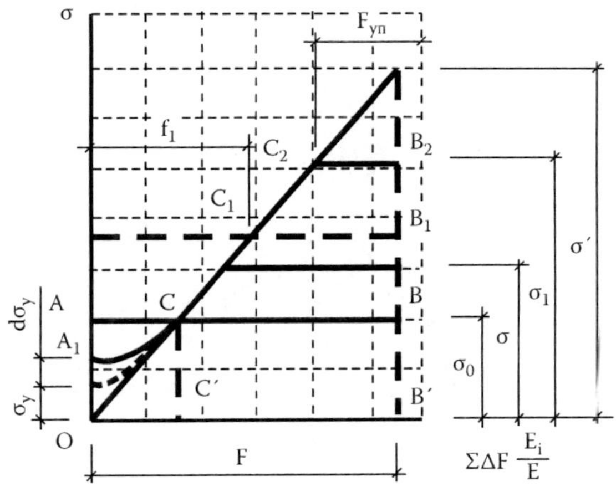

FIGURE 2.44 Structural diagram of nonlinearly polymeric body. (Reprinted from Yu. Potapov, O. Figovsky, Yu. Borisov, S. Pinaev, and D. Beilin, "Creep of Polymer Concrete at Compressive Loading," *J. Scientific Israel Technological Advantages* 5, nos. 1–2 (2003): 1–10. With permission.)

Let F be the total area of the cross section of a polymeric system and F_e the area of elastic elements of system. Then, from Figure 2.44: $\sigma_0 F = 0.5\, \sigma_1 (F + F_e)$ where: $F_e = \sigma' - \sigma_1/\sigma'$ and $\sigma_1 = \varepsilon E$. Consequently, $\sigma_0 = \varepsilon E\,(0.5 + (\sigma' - \varepsilon E/2\sigma')$. Hence, it follows that

$$\sigma_0 = \varepsilon E\,(1 - \varepsilon E/2\sigma'). \tag{2.45}$$

Because Formula (2.45) is identical to the experimentally obtained Formula (2.41), it is obvious that the structural diagram (Figure 2.44) adequately describes the behavior of rubber concretes at compression and can be added to the basis of the creep equation of these materials. (The similar diagram corresponds to a number of plastic [rigid PVC, polyethylene] and furan polymeric concrete.)

Let's consider the process of creep at compression of RubCon samples under its structural diagram (Figure 2.44). The initial loading is represented by the straight line AB. Thus, all particles of RubCon composition are divided into two kinds: particles for which the limit of long-term strength is surpassed, (line segment) OC' and particles for which the limit of long-term strength is not surpassed.

Let's establish an increment of deformation and, hence, stress $d\sigma = d\varepsilon E$. For the moment of time dt, the superfluous stress will decrease due to its distribution over a greater number of particles. It is obvious that the average value of an increment of viscous highly elastic deformations of particles coincides with the average value of elastic deformation. Further, we shall designate superfluous force $\sigma_e f_l$ and notice that for an interval of time dt, stress is constant. In this case, Eq. (2.44) will become:

$$(d\varepsilon_n/dt)\, f_1 = \sigma_e f/\eta_1 + \sigma_e f/\eta_1 \,(\exp(-E_1\, t/\eta)) \tag{2.46}$$

where a_n is the viscous, highly elastic creep deformation. Reduction of superfluous force is equal to the increase in elastic force, that is

$$d\sigma_{e\,*} f = d\sigma * (F - f_1),$$

where

$$d\sigma_{e\,*} f_1/E = d\sigma/E*(F - f_1),$$

or

$$d\varepsilon_n * f_1 = d\varepsilon\,(F - f_1) \tag{2.47}$$

From the structural diagram we find:

$$F - f_1 = F(1 - \sigma/\sigma'). \tag{2.48}$$

The superfluous force is connected with active force with the relationship

$$\sigma_{e\,*} f_1 = \sigma_{0\,*} F - \sigma\,(1 - \sigma/2\sigma'). \tag{2.49}$$

Substituting Equations (2.47), (2.48), and (2.49) in (2.46), we obtain:

$$(d\varepsilon_n/dt)*(1 - \sigma/\sigma')F = (\sigma_0 - \sigma\,(1 - \sigma/2\sigma'))*F(1/\eta_1 + (1/\eta)*(\exp(-E_1\, t/\eta))).$$

Taking into account that $\sigma = \varepsilon E$ we have

$$(1/E)*(d\sigma\,(1 - \sigma/\sigma'))/\sigma_0 - \sigma\,(1 - \sigma/2\sigma') = dt/\eta_1 + (dt/\eta)*(\exp(-E_1\, t/\eta)).$$

After integration at the initial conditions $t = 0$ and $\sigma_0 = \sigma$ we shall have

$$\varepsilon\,(1 - E\varepsilon/2\sigma') = \varepsilon_0\,(1 - (\varepsilon_0/2\sigma')\exp - ((Et/\eta_1) + (E/E_1)(1 - \exp(-E_1t/\eta))).\quad(2.50)$$

Let's estimate the adequacy of Equation (2.50) in the results of the experiments in Figures 2.39 and 2.40. With this purpose, we shall create the structural diagram (Figure 2.45) similar the one shown in Figure 2.44. Figure 2.45 shows that structural diagrams are a little bit displaced from the beginning coordinates and on an axis of ordinates cut a piece $c = 0.06$. In this connection, Equation (2.50) is transformed:

$$(\varepsilon + c)(1 - (\varepsilon + c/(2\sigma'/E)) = (\varepsilon_0 + c)(1 - ((\varepsilon_0 + c)/2\sigma')\exp - ((Et/\eta_1) + (E/E_1)(1 - \exp(-E_1t/\eta)).\quad(2.51)$$

Let's perform the linearization of Equation (2.51). For this purpose, the following transformation is feasible:

$$((\varepsilon_0 + c) - (\varepsilon + c)(1 - (\varepsilon + c)/2\varepsilon'))/(\varepsilon_0 + c)^2/2\varepsilon' = \exp - ((Et/\eta_1) + (E/E_1)(1 - \exp(-E_1t/\eta))).\quad(2.52)$$

Having designated the left part through Y, after taking the logarithm we have

$$\ln Y = (Et/\eta_1) + (E/E_1)(1 - \exp(-E_1t/\eta))\quad(2.53)$$

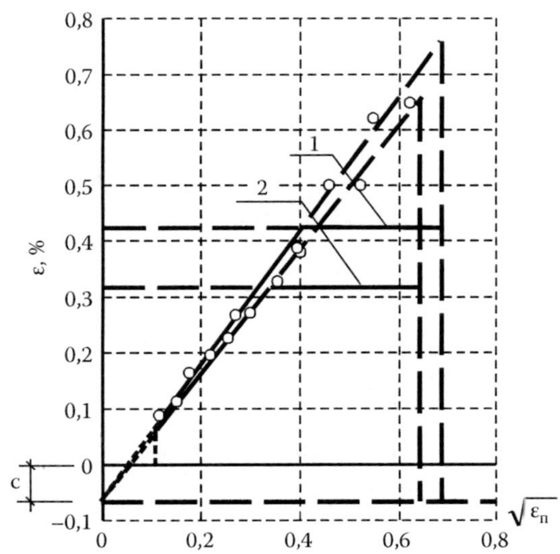

FIGURE 2.45 Structural diagram of RubCon samples at compression: (1) samples type A, (2) samples type B. (Reprinted from Yu. Potapov, O. Figovsky, Yu. Borisov, S. Pinaev, and D. Beilin, "Creep of Polymer Concrete at Compressive Loading," *J. Scientific Israel Technological Advantages* 5, nos. 1–2 (2003): 1–10. With permission.)

At $t \to \infty$ from (2.53) we shall receive the expression for asymptote:

$$\ln Y = (-E_1/\eta t) - E/E_1). \tag{2.54}$$

At $t = 0$ we have the relationship of the modulus of elasticity with the modulus of high-elasticity phases:

$$E/E_1 = \ln Y. \tag{2.55}$$

To find the relationship of the high-elasticity phase's modulus to the coefficient of internal friction, we shall write (2.55) as

$$1 - (\ln Y + Et/\eta_1)/(E/E_1) = \exp(-E_1 t/\eta). \tag{2.56}$$

Having designated the left part through X, after taking the logarithm we have

$$\ln X = -E_1 t/\eta \tag{2.57}$$

Value E_1/η can be determined as a tangent of the straight line inclination of diagram $\ln X - t$ (time) to an abscissa t.

Diagrams of the creep curves are shown in Figures 2.46 and 2.47, constructed using Formulas (2.53) and (2.57), respectively; in so doing, structural coefficient ε' was calculated using the formula $2\varepsilon' = (\varepsilon + c)^2/\varepsilon_n$. The values of the coefficients obtained from the structural diagram (Figure 2.45) were equal: $2\varepsilon' = 1.36\%$ and $c = 0.06\%$ (sample no. 5 of type A); $2\varepsilon' = 1.31\%$ and $c = 0.06\%$ (sample no. 6 of type B).

Linearized diagrams of creep of the RubCon samples, constructed using Equation (2.51) show good agreement with the experimental results. The obtained experimental values of creep are given in Table 2.6. We can also note that at an identical loading level of RubCon samples based on the type A liquid rubber, the relaxation of a viscous phase proceeds faster than in the samples based on type B rubber. This fact highlights the significant plasticity of type B rubber, which has a high strength as well. It is obvious that at design of RubCon type B load-bearing structures at long-term loading action, creep deformations will be large, which will have an effect on redistribution of forces in statically indeterminate structures.

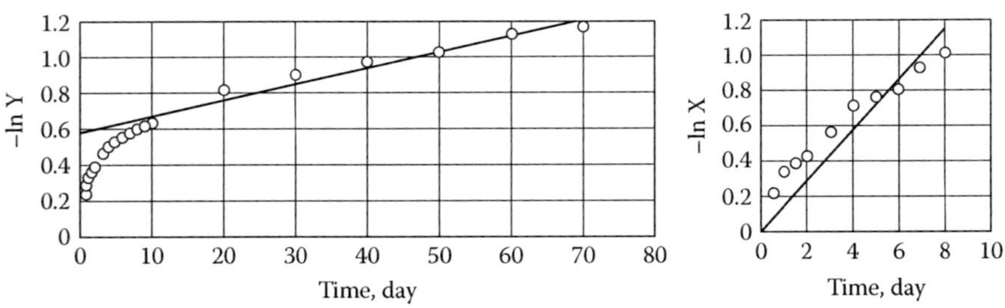

FIGURE 2.46 Diagrams of creep after linearization (sample no. 5 of type A). (Reprinted from Yu. Potapov, O. Figovsky, Yu. Borisov, S. Pinaev, and D. Beilin, "Creep of Polymer Concrete at Compressive Loading," *J. Scientific Israel Technological Advantages* 5, nos. 1–2 (2003): 1–10. With permission.)

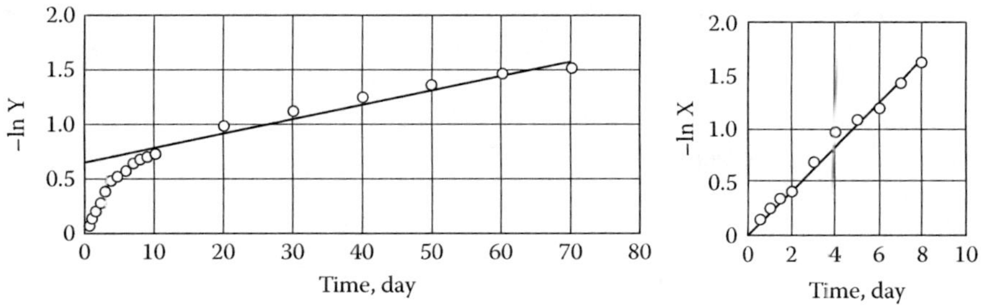

FIGURE 2.47 Diagrams of creep after linearization (sample no. 6 of type B). (Reprinted from Yu. Potapov, O. Figovsky, Yu. Borisov, S. Pinaev, and D. Beilin, "Creep of Polymer Concrete at Compressive Loading," *J. Scientific Israel Technological Advantages* 5, nos. 1–2 (2003): 1–10. With permission.)

TABLE 2.6
Time of Relaxation of RubCon Samples at Compression

Type of Samples	Number of Samples	$\frac{E}{E_1}$	$\frac{E}{\eta_1}$ (day)$^{-1}$	$\frac{E_1}{\eta}$ (day)$^{-1}$	Time of Relaxation $T_1 = \frac{E}{\eta_1}$ (day)	Time of Relaxation $T_2 = \frac{E_1}{\eta}$ (day)	$\frac{T_1}{T_2}$
A	9	–0.770	0.222	–0.080	4.5	12.5	0.36
	8	–0.746	0.175	–0.145	5.7	6.9	0.83
	7	–0.690	0.161	–0.199	6.2	5.0	1.24
	6	–0.650	0.080	–0.337	12.5	2.9	4.21
	5	–0.610	0.004	–0.440	227.3	2.3	100
B	9	–0.918	0.113	–0.080	8.9	12.5	0.71
	8	–0.215	0.069	–0.145	14.5	6.9	2.10
	7	–0.140	0.010	–0.199	105.3	5.0	20.95
	6	–0.139	0.010	–0.445	100.0	2.2	45.41
	5	–0.130	0.007	–0.650	142.8	1.5	92.86

Source: Reprinted from Yu. Potapov, O. Figovsky, Yu. Borisov, S. Pinaev, and D. Beilin, "Creep of Polymer Concrete at Compressive Loading," *J. Scientific Israel Technological Advantages* 5, nos. 1–2 (2003): 1–10. With permission.

Creep Deformation of Plain RubCon with a Combination of Long-Term Compressive Loading and an Aggressive Environment

In a situation that combines long-term compressive load and an aggressive environment, the mechanical properties of a material deteriorate because the complex influence of these factors is accompanied by a synergetic effect. Research on RubCon creep in these conditions was carried out on samples measuring 40 × 40 × 160 mm in the special chamber at a temperature of 20°C±2°C. Three series of samples were tested in different corrosion environments: in water, in a 30% solution of a sulfuric acid, and in a 10% solution of a caustic sodium. Each series is composed of 7 samples loaded on 80%, 75%, 70%, 60%, 59%, 40%, and 30% of ultimate strength, determined from short-term tests on the control samples.

Results of these experiments are shown in Figures 2.48–2.50. The obtained experimental data creep in RubCon samples were analyzed with the phenomenological theory of structural diagrams (Figure 2.51a–c).

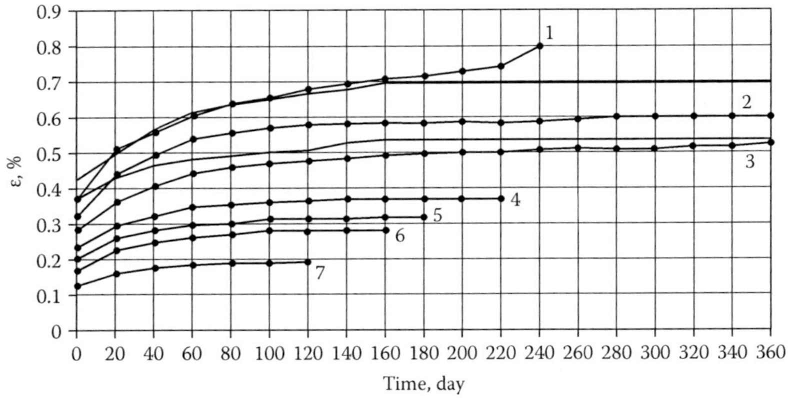

FIGURE 2.48 Creep curves of *RubCon* samples in water environment: (1) σ_0 = 73.9 MPa; ε_0 = 0.346%; (2) σ_0 = 69.5 MPa; ε_0 = 0.315%; (3) σ_0 = 64.7 MPa; ε_0 = 0.275%; (4) σ_0 = 55.5MPa; ε_0 = 0.219%; (5) σ_0 = 46.2 MPa; ε_0 = 0.198%; (6) σ_0 = 36.9 MPa; ε_0 = 0.173%; (7) σ_0 = 27.7 MPa; ε_0 = 0.13. (Reprinted from Yu. Potapov, O. Figovsky, Yu. Borisov, S. Pinaev, and D. Beilin, "Creep of Polymer Concrete at Compressive Loading," *J. Scientific Israel Technological Advantages* 5, nos. 1–2 (2003): 1–10. With permission.)

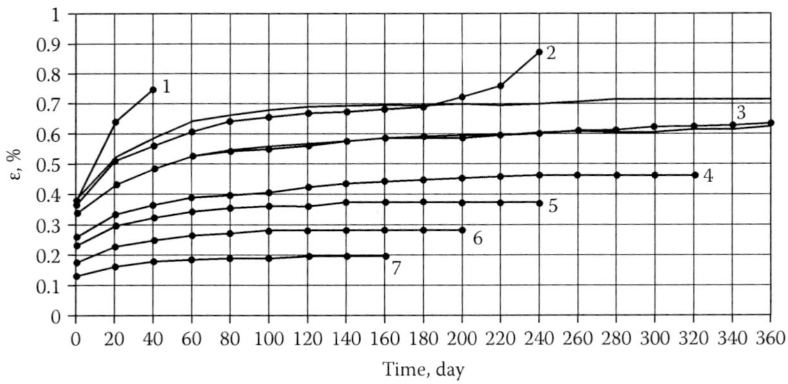

FIGURE 2.49 Creep curves of *RubCon* samples in 30% solution of sulfuric acid: (1) σ_0 = 77.12 MPa; ε_0 = 0.35%; (2) σ_0 = 72.3 MPa; ε_0 = 0.320%; (3) σ_0 = 67.48 MPa; ε_0 = 0.305%; (4) σ_0 = 57.84 MPa; ε_0 = 0.254%; (5) σ_0 = 48.2 MPa; ε_0 = 0.219%; (6) σ_0 = 38.56 MPa; ε_0 = 0.176%; (7) σ_0 = 28.92 MPa; ε_0 = 0.127. (Reprinted from Yu. Potapov, O. Figovsky, Yu. Borisov, S. Pinaev, and D. Beilin, "Creep of Polymer Concrete at Compressive Loading," *J. Scientific Israel Technological Advantages* 5, nos. 1–2 (2003): 1–10. With permission.)

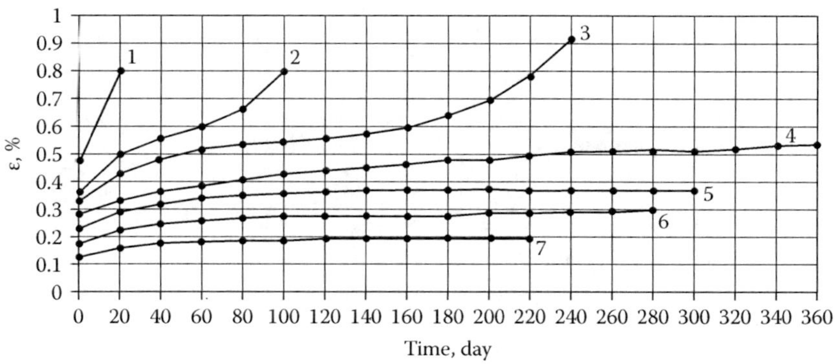

FIGURE 2.50 Creep curves of *RubCon* samples in 10% solution of solution of a caustic sodium: (1) $\sigma_0 = 75.44$ MPa; $\varepsilon_0 = 0.368\%$; (2) $\sigma_0 = 70.73$ MPa; $\varepsilon_0 = 0.336\%$; (3) $\sigma_0 = 66.01$ MPa; $\varepsilon_0 = 0.293\%$; (4) $\sigma_0 = 56.68$ MPa; $\varepsilon_0 = 0.259\%$; (5) $\sigma_0 = 47.15$ MPa; $\varepsilon_0 = 0.212\%$; (6) $\sigma_0 = 37.72$ MPa; $\varepsilon_0 = 0.184\%$; (7) $\sigma_0 = 28.29$ MPa; $\varepsilon_0 = 0.135$. (Reprinted from Yu. Potapov, O. Figovsky, Yu. Borisov, S. Pinaev, and D. Beilin, "Creep of Polymer Concrete at Compressive Loading," *J. Scientific Israel Technological Advantages* 5, nos. 1–2 (2003): 1–10. With permission.)

The analysis of structural diagrams shows that at creep in aggressive environments, the extent of a viscous phase in the composite structure is increased due to penetration of the environment into the material pores and its influence on the structure and rheological properties of RubCon. The increase of a viscous phase results in a decrease in strength and deformation characteristics of RubCon.

Analysis of the experiment has shown a combination of long-term compressive loading and the aggressive environment coefficient of chemical resistance is: k = 0.76 for water, k = 0.71 for a 30% solution of sulfuric acid, and k = 0.65 for a 10% solution of caustic sodium. Thus, it is possible to conclude that RubCon keeps high strength at complex influence of the aggressive environment and external loading.

Creep Deformation of Fibrous RubCon at Long-Term Compressive Load

Resistance of fiber-reinforced RubCon to long-term loading is the principal criterion of its application as a structural material. We investigated creep of plain RubCon at compression earlier. Experiments were continued for the purpose of studying fibrous RubCon creep.

Samples measuring 40 × 40 × 160 mm were made using steel fibers with copper covering. The reinforcement ratio μ = 2%, aspect ratio L/d = 100, and maximum coarse aggregate particle size a = 5 mm conformed to the values obtained above.

Two series of fibrous RubCon samples were tested. Samples of series A contained a fine aggregate (sand); samples from series B contained coarse aggregate (chippings) and fine (sand) aggregates. Samples were exposed to compressive force, which corresponded to compressive stress (σ_c) equal to 85%, 80%, 75%, 70%, 60%, 50%, 40%, 30%, and 20% of ultimate compressive strength (σ_{uc}) at short-term loading of control samples. Ultimate compressive strengths were σ_{uc} = 112 MPa and σ_{uc} = 107 MPa for series A and B samples, respectively. Curves of creep are plotted in Figures 2.52 and 2.53.

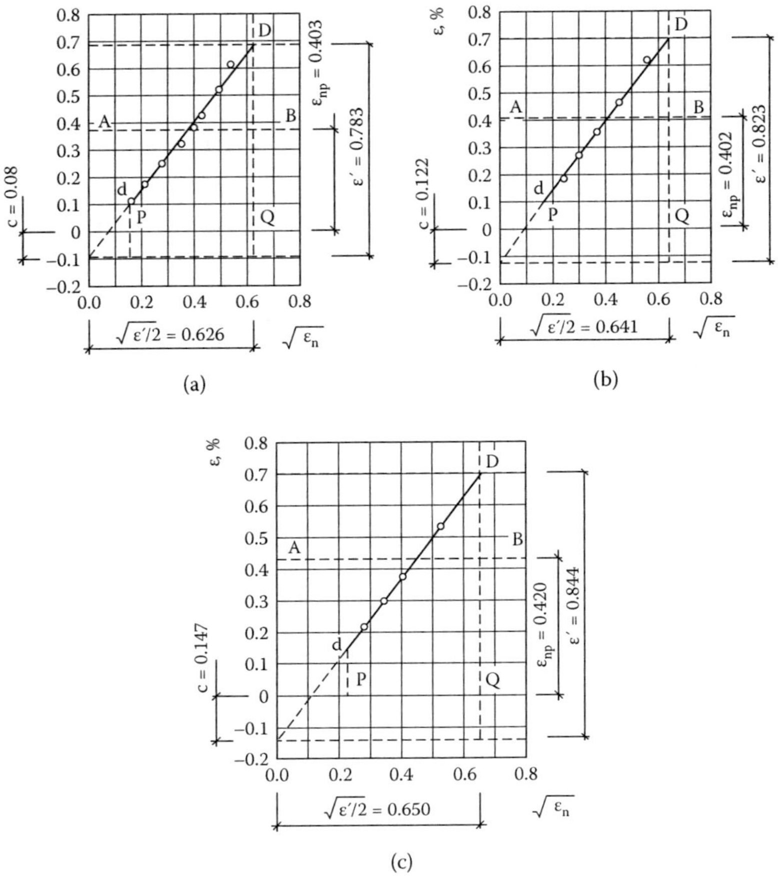

FIGURE 2.51 Structural diagram of RubCon samples: (a) in water environment, (b) in 30% solution of sulfuric acid, (c) in 10% solution of solution of a caustic sodium. (Reprinted from Yu. Potapov, O. Figovsky, Yu. Borisov, S. Pinaev, and D. Beilin, "Creep of Polymer Concrete at Compressive Loading," *J. Scientific Israel Technological Advantages* 5, nos. 1–2 (2003): 1–10. With permission.)

It can be seen that creep deformations are damped out with time at compressive stress $\sigma_c < 0.7\sigma_{uc}$.

To gain a better understanding of the influence of fiber reinforcement on creep of RubCon, we investigated the relationship between values of compressive stresses and creep deformations that are damped out with time. The analysis of the result diagrams (Figures 2.54 and 2.55) shows linear stress–strain dependence of fiber-reinforced and plain RubCon samples at short-term compressive loading. However, creep deformations of these samples do not linearly depend on compressive stress value due to highly elastic deformation of the polybutadiene binder.

The allowed us to obtain RubCon creep coefficients $k_{creep} = \sigma_{creep}/\sigma_{uc}$. The relationship of creep coefficients to time for the reinforced and plain RubCon samples is shown in Figure 2.56.

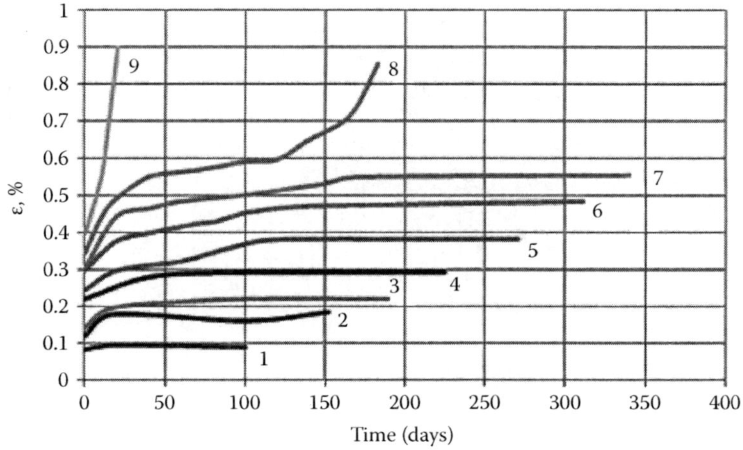

FIGURE 2.52 Creep curves of fiber reinforced RubCon contained fine aggregate (sand): (1) $\sigma_c = 22.4$ MPa, $\varepsilon_c = 0.095\%$; (2) $\sigma_c = 33.6$ MPa, $\varepsilon_c = 0.15\%$; (3) $\sigma_c = 44.8$ MPa, $\varepsilon_c = 0.20\%$; (4) $\sigma_c = 56$ MPa, $\varepsilon_c = 0.28\%$; (5) $\sigma_c = 67.2$ MPa, $\varepsilon_c = 0.33\%$; (6) $\sigma_c = 78.4$ MPa, $\varepsilon_c = 0.37\%$; (7) $\sigma_c = 84$ MPa, $\varepsilon_c = 0.402\%$; (8) $\sigma_c = 89.6$ MPa, $\varepsilon_c = 0.43\%$; (9) $\sigma_c = 95.2$ MPa, $\varepsilon_c = 0.45\%$. (Reprinted from Yu. Potapov, O. Figovsky, Yu. Borisov, S. Pinaev, and D. Beilin, "Creep of Polymer Concrete at Compressive Loading," *J. Scientific Israel Technological Advantages* 5, nos. 1–2 (2003): 1–10. With permission.)

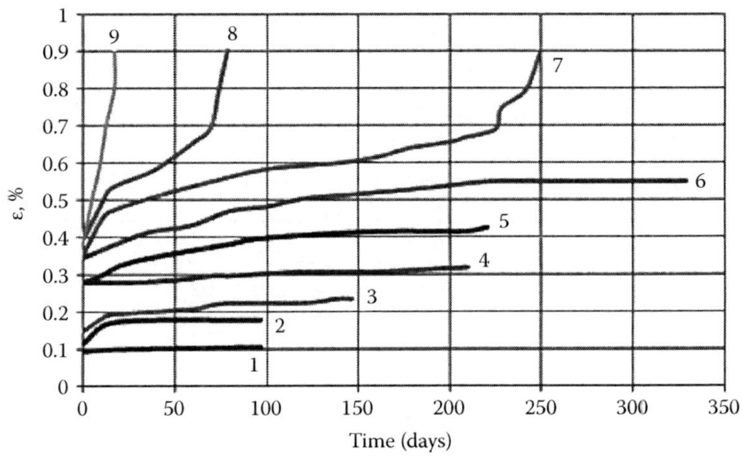

FIGURE 2.53 Creep curves of fiber reinforced RubCon contained fine (sand) and coarse (chippings) aggregates: (1) $\sigma_c = 21.4$ MPa, $\varepsilon_c = 0.078\%$; (2) $\sigma_c = 32.1$ MPa, $\varepsilon_c = 0.117\%$; (3) $\sigma_c = 42.8$ MPa, $\varepsilon_c = 0.153\%$; (4) $\sigma_c = 53.5$ MPa, $\varepsilon_c = 0.205\%$; (5) $\sigma_c = 64.2$ MPa, $\varepsilon_c = 0.26\%$; (6) $\sigma_c = 74.9$ MPa, $\varepsilon_c = 0.309\%$; (7) $\sigma_c = 80.3$ MPa, $\varepsilon_c = 0.34\%$; (8) $\sigma_c = 85.6$ MPa, $\varepsilon_c = 0.37\%$; (9) $\sigma_c = 90.92$ MPa, $\varepsilon_c = 0.41\%$. (Reprinted from Yu. Potapov, O. Figovsky, Yu. Borisov, S. Pinaev, and D. Beilin, "Creep of Polymer Concrete at Compressive Loading," *J. Scientific Israel Technological Advantages* 5, nos. 1–2 (2003): 1–10. With permission.)

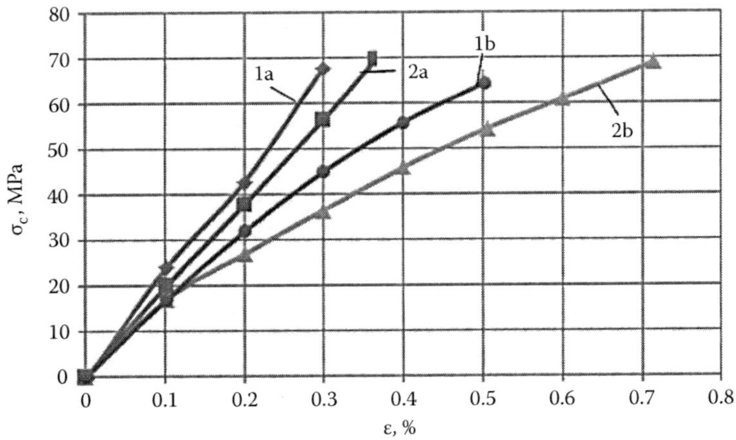

FIGURE 2.54 Stress–strain relationship of *plain* RubCon samples at short-term (1) and long-term (2) compressive loading. (a) Samples contained fine and coarse aggregates, (b) samples contained fine aggregate. (Reprinted from Yu. Potapov, O. Figovsky, Yu. Borisov, S. Pinaev, and D. Beilin, "Creep of Polymer Concrete at Compressive Loading," *J. Scientific Israel Technological Advantages* 5, nos. 1–2 (2003): 1–10. With permission.)

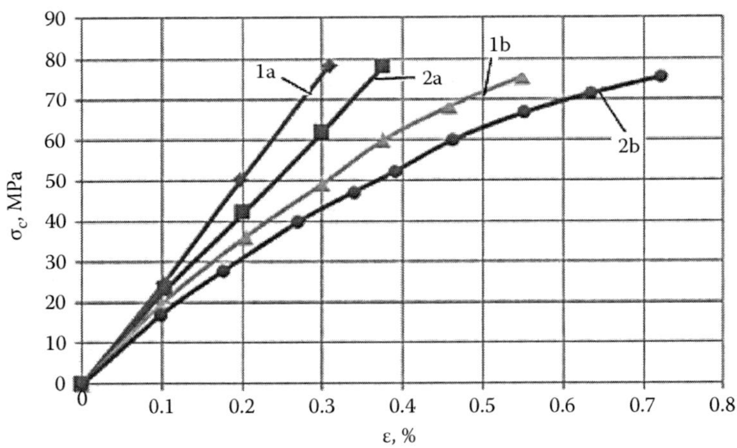

FIGURE 2.55 Stress–strain relationship of *fiber reinforced RubCon* samples at short-term (1) and long-term (2) compressive loading. (a) Samples contained fine and coarse aggregates, (b) samples contained fine aggregate. (Reprinted from Yu. Potapov, O. Figovsky, Yu. Borisov, S. Pinaev, and D. Beilin, "Creep of Polymer Concrete at Compressive Loading," *J. Scientific Israel Technological Advantages* 5, nos. 1–2 (2003): 1–10. With permission.)

Asymptotes of the curves give an indication of the fibrous RubCon creep limit. It is interesting to note that the creep coefficient of the plain RubCon k_{creep} = 0.77–0.78, whereas for fiber reinforced RubCon it is slightly below k_{creep} = 0.74–0.75. In this case the creep limit of fiber-reinforced RubCon at compression σ_{creep} = 74.9–78.4 MPa exceeds the creep limit of plain material σ_{creep} = 66–66.8 MPa. This phenomenon is

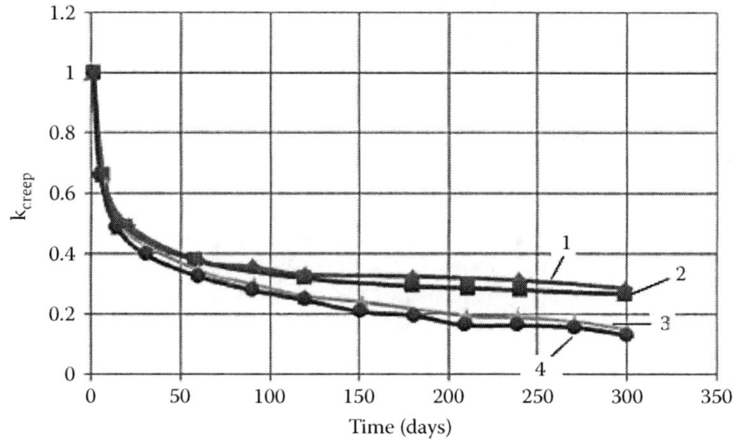

FIGURE 2.56 Dependence of creep coefficient on time: (1) plain RubCon samples contained fine and coarse aggregates, (2) plain RubCon samples contained fine aggregate, (3) fiber-reinforced RubCon samples contained fine and coarse aggregates, (4) fiber-reinforced RubCon samples contained fine aggregate. (Reprinted from Yu. Potapov, O. Figovsky, Yu. Borisov, S. Pinaev, and D. Beilin, "Creep of Polymer Concrete at Compressive Loading," *J. Scientific Israel Technological Advantages* 5, nos. 1–2 (2003): 1–10. With permission.)

associated with highly elastic deformation of the polybutadiene matrix because the creep stress of the fibrous RubCon matrix is more than the creep stress of RubCon itself.

The creep limit $\sigma_{ult.creep}$ and elastic modulus E_{creep} of fiber-reinforced *RubCon* at long-term compressive loading were obtained experimentally. For RubCon containing fine aggregate: $\sigma_{ult.\ creep} = 78.4$ MPa, $E_{creep} = 20300$ MPa, for RubCon containing fine and coarse aggregate: $\sigma_{ult.\ creep} = 74.9$ MPa, $E_{creep} = 21000$ MPa.

Summary

- Creep deformation of rubber concretes has a damping character.
- The mathematical model of creep deformations at long-term compressive loading agrees well with experimental results and can be put in a basis of design of RubCon load-bearing structures.
- Steel fibrous RubCon is a creep-resistant material; the coefficient of creep at compressive load $k_{cr} \approx 0.75$.

CHEMICAL RESISTANCE OF RUBCON [1,3,21–28]

Corrosive attack of an aggressive medium on composite polymeric materials is manifested as a change in its structure and properties without disruption of integrity or with destruction of the materials. The corrosion environment penetrates between macromolecules through micropores and fine capillaries and irreversibly changes the chemical structure of the composite and reinforcing bars or fibers as well. Such changes of RubCon structure are characterized by formation of new active groups in molecular chains (carboxyl, hydroxyl, amine, ketone) and the presence of double links resulting in a lowering of physical–mechanical properties of the material. Water and water solutions of electrolytes adversely affect composite materials, causing their destruction, cracking, and so on.

As a rule, the most widespread liquid environments are water solutions of acids, alkalis, and salts. Chemical resistance of RubCon was specified as a change in weight and compressive strength of the samples after their exposure to aggressive liquid environments during a certain period of time. The choice of aggressive environment compositions was based on their prevalence in industrial production.

Coefficient of Chemical Resistance of RubCon

A basic guideline in a choice of corrosion environments during the test of RubCon specimens was their wide spreading into industrial production. Such environments were water, 30% and 70% solutions of sulfuric acid, 5% solutions of phosphoric and acetic acids, 3% solution of nitric acid, 3% and 30% solutions of hydrochloric acid, 10% solutions of lactic and lemon acids, caustic soda and caustic potash, diesel fuel, acetone, 25% water solution of ammonia, 30% solution of copper vitriol, and a saturated solution of sodium chloride. Chemical resistance of RubCon was estimated on test specimens measuring 4 × 4 × 16 cm [21–23].

The following values were determined in the course of experiments before and after attack by water and corrosive media during one year of exposure at room temperature:

- Δm = change of test specimen mass
- R_c = compressive strength
- E = modulus of elasticity
- K_{cr}- = coefficient of chemical resistance
- K_E = coefficient of change of modulus of elasticity
- x = depth of penetration of corrosive liquid (obtained by microscope-assisted measurement of the visible front of diffusion liquid movement).

It is obvious that the durability of a structural material depends on its resistance in a corrosive environment. Indexes of this resistance are as follows:

$$\text{coefficient of chemical resistance: } K_{\text{cr}} = R_{c1}/R_{c0}, \tag{2.58}$$

$$\text{coefficient of change of the modulus of elasticity: } K_E = E_1/E_0. \tag{2.59}$$

$$\text{change of test specimen mass: } \Delta m = (m_1 - m_0)*100/m_0 \tag{2.60}$$

Here, R_{c0}, E_0, m_0, R_{c1}, E_1, and m_1 are compressive strength, modulus of elasticity, and mass of the specimen, respectively, before and after corrosive media attack during one year of exposure.

TABLE 2.7
Physical–Mechanical and Chemical Properties of RubCon after Attack of Corrosive Environments

	Indexes			
Corrosive Media	Δm,%	K_{cr}	K_E	x, мм
Water	0.05	1.00	1,00	—
30% solution of sulfuric acid	0.28	0.95	0.87	1.33
70% solution of sulfuric acid	0.33	0.92	0.90	1.12
5% solution of phosphoric acid	0.14	0.94	0.81	0.93
3% solution of nitric acid	0.63	0.80	0.70	3.40
5% solution of hydrochloric acid	0.13	0.81	0.81	0.81
36% solution of hydrochloric acid	1.14	0.69	0.53	5.10
5% solution of acetic acids	0.22	0.82	0.78	1.32
10% solution of lactic acids	0.28	0.95	0.93	1.57
10% solution of lemon acids	0.16	0.87	0.86	0.92
25% water solution of ammonia	0.31	0.82	0.70	1.35
10% solution of caustic soda	0.22	0.87	0.77	1.15
10% solution of caustic potash	0.17	0.91	0.85	0.92
Diesel fuel	0.28	0.88	0.86	2.25
Acetone	0.25	0.88	0.88	1.77
30% solution of copper vitriol	0.22	0.84	0.81	1.32
Saturated solution of sodium chloride	0.16	0.96	0.95	0.92

Source: Reprinted from V. Solomatov and V. Seljaev, "Chemical Strength of Composite Building Materials," *Stroiizdat*, Moscow, 1987 (in Russian). With permission.

The results of the experiments are given in Table 2.7 [21].

As shown in Table 2.7, the coefficient of chemical resistance of RubCon is 1.0 for water, 0.81–0.95 for all mineral acids (exception is K_{cr} = 0.69 for 36% solution of hydrochloric acid), 0.82–0.95 for organic acids, 0.82–0.91 for alkalis, 0.88 for solvents and petroleum products, and 0.84–0.86 for solution of salts. The analysis of experimental data has shown that RubCon offers universal chemical resistance many times higher than ordinary Portland cement concrete and surpasses the chemical resistance of polyester, polyepoxy, and vinyl ester polymer concrete. It is worth noting that penetration depths of 5% nitric and 36% solution of hydrochloric acids into RubCon sample bodies were 3, 4, and 5.1 mm, respectively; the penetration ability of these acids is higher as compared with other corrosive environments.

The dynamics of changes in compressive strength of RubCon samples during corrosion medium exposure is shown in Figure 2.57.

It is shown [18] that RubCon has high chemical resistance not only under normal operating conditions, but also at long-term compressive loading (creep) in aggressive environments.

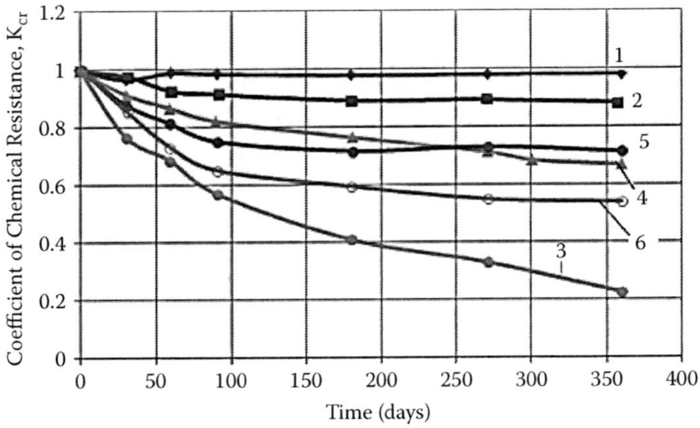

FIGURE 2.57 Relationship between the RubCon coefficient of chemical resistance and time of corrosion medium exposition: (1) water, (2) 30% solution of sulfuric acid, (3) 36% solution of hydrochloric acid, (4) 10% solution of caustic soda, (5) diesel fuel, (6) 5% solution of acetic acids. (Reprinted from Yu. Potapov, Yu. Borisov, V. Chmyhv, and D. Beilin, "Research of Polymer Concrete Based on Low Molecular Polybutadiene, Part VIII: Chemical Resistance of Polymer Concrete," *J. Scientific Israel Technological Advantages* 7, nos. 1–2 (2005): 72–78. With permission.)

Corrosive Environment Influence on Concrete with a Polybutadiene Matrix

The physical-chemical impact of a corrosive environment on a polymer composite material is manifested in changes in its structure, sometimes resulting in destruction [3]. Figure 2.57 shows that the most intensive decrease of RubCon strength occurs during the first 6 months of exposure to a corrosive environments. Penetration of the corrosive environment between macromolecules of a material through its micropores and fine capillaries produces a reduction of surface energy on the body–environment boundaries. This leads to formation and development of cracks in a composite, lowering of its strength, elevation of polymeric chain flexibility, and plasticization of a material.

Destruction of polybutadiene that occurs under the action of nitric acid is caused by oxidation of a polymer macromolecule. In other words, cross-section links of a spatial composite network formed by the vulcanization process are broken. The well-known oxidization ability of sulfuric acid is responsible for a decrease in double links in the rubber molecular structure, resulting in a reduction in the RubCon strength indexes. Corrosive attack by hydrochloric acid is linked to oxidation and isomerization processes and, therefore, the durability of the composite depends on the speed of these processes.

Corrosive attack by organic carboxylic acids on polybutadiene binder is exhibited by oxidization and depolymerization processes, with the resulting the formation of carboxyl groups COOH. Carboxylic acids generally are weaker than inorganic acids because the speed of these processes is lower and, therefore, the decrease of physical–mechanical properties of RubCon occurs more slowly. The most intensive reduction of RubCon strength properties (compressive strength and modulus of elasticity) due to carboxylic acids attack were noted in the first 90 days. Penetration depths of corrosive medium into the body of test specimens after one year of exposure are 1, 3, 1.6, and 0.9 mm for acetic, lactic, an lemon acids, respectively (Table 2.7).

Table 2.7 shows that among carboxylic acids, acetic acid is the most aggressive medium relative to RubCon because it is the strongest oxidizer. Lactic and lemon acids are less aggressive.

Temperature Influence on Chemical Resistance of RubCon

The behavior of RubCon at elevated and negative temperatures in nonaggressive environmental operating conditions have been studied [24]. Continuation of this research involves the examination of RubCon resistance in liquid corrosive environments at elevated temperatures, which showed visual destruction of the material. The combined influence of temperature and corrosive medium was determined by testing RubCon specimens measuring 4 × 4 × 16 cm. Experiments were performed in a 10% solution of caustic soda and a 30% solution of sulfuric acid at temperatures of 40°C and 60°C. Results of the experiments are illustrated in Table 2.8.

It can be seen that the elevated temperature reduces the chemical resistance of RubCon samples and leads to their destruction due to accelerated chemical reactions between the corrosion environment and the material.

Forecast of Chemical Resistance Coefficient

The forecast of RubCon chemical resistance coefficient for τ = 10 years of operation in corrosive environments has been carried out by [21,24] according to these formulas:

$$\lg K_{cr} = a + b \lg \tau \tag{2.61}$$

TABLE 2.8
Impact of Elevated Temperatures on Chemical Resistance of RubCon

		Indexes	
Corrosive Media	**`T°C**	**Δm,%**	**K_{cr}**
Water	+20	0.04	0.99
	+40	0.06	0.99
	+60	0.07	0.98
30% solution of sulfuric acid	+20	0.16	0.97
	+40	0.20	0.95
	+60	0.23	0.93
10% solutions of caustic soda	+20	0.14	0.92
	+40	0.16	0.90
	+60	0.19	0.87

Source: Reprinted from V. Solomatov and V. Seljaev, "Chemical Strength of Composite Building Materials," *Stroiizdat*, Moscow, 1987 (in Russian). With permission.

TABLE 2.9
Predicted Chemical Resistance of *RubCon* after 10 Years of Operation in Corrosive Environments

Corrosive Media	Predicted Values of K_{cr}
30% solution of sulfuric acid	0.914
70% solution of sulfuric acid	0.865
5% solution of phosphoric acid	0.892
3% solution of nitric acid	0.682
5% solution of hydrochloric acid	0.706
36% solution of hydrochloric acid	0.521
5% solution of acetic acids	0.728
10% solution of lactic acids	0.923
10% solution of lemon acids	0.799
25% water solution of ammonia	0.724
10% solution of caustic soda	0.800
10% solution of caustic potash	0.937
Diesel fuel	0.800
Acetone	0.820
30% solution of copper vitriol	0.737
Saturated solution of sodium chloride	0.927

Source: Reprinted from V. Solomatov and V. Seljaev, "Chemical Strength of Composite Building Materials," *Stroiizdat*, Moscow, 1987 (in Russian). With permission.

$$a = \lg K' - b \lg \tau';$$

$$b = [\Sigma^n_{i=1} (\lg K' - \lg K_i)(\lg \tau' - \lg \tau_i)/\Sigma^n_{i=1} (\lg \tau' - \lg \tau_i)];$$

$$\lg K' = \Sigma^n_{i=1} \lg K_i/n;$$

$$\lg \tau' = \Sigma^n_{i=1} \lg \tau_i/n$$

where K_i, τ_i are the coefficients of chemical resistance and intervening testing durations (30, 60, 90, 180, and 270 days), respectively. Results of calculations are shown in Table 2.9.

It is evident that RubCon is a corrosive-stable structural material because its chemical resistance coefficient during 10 years of operation in aggressive environments exceeds 0.5. The obtained data allow determination of the thickness of the concrete cover of a reinforced RubCon structure.

Enhancement of Chemical Resistance of RubCon

As follows from Table 2.7, the RubCon coefficient of chemical resistance in a 36% solution of hydrochloric acid is minimal. We launched the series of experiments to raise the RubCon chemical resistance by introduction of special active additives in the vulcanization process. These additives make it possible to increase the density of spatial cross links of polymeric molecules and, by doing so, improve the physical–mechanical properties and resistibility of RubCon to corrosive medium exposure [21,23].

The additives were thiurarn-D and Altax®. Optimum quantity of the additives in the RubCon composition was obtained by a two-factor experiment. In this case, the dosage of the additives was the varied parameter and the compressive strength of RubCon test specimens was the efficiency function. Figure 2.58 shows the efficiency function surface for RubCon samples in a 36% solution of hydrochloric acid, depending on the quantity of additives: sulfur + thiurarn-D (x_1) and Altax (x_2).

The regression formula for the relationship between RubCon compressive strength R and quantities of additives (x_1 = sulfur + thiurarn-D and x_2 = Altax) is

$$R = -369.5 + 162.7\,x_1 + 774.5\,x_2 - 15.5\,x_1^2 - 553.2\,x_2^2 - 106.0\,x_1 x_2. \quad (2.62)$$

The compressive strength of the RubCon samples after 90 days of exposure in a 36% solution of HCl peaks at x_1 = 4.45% and x_2 = 0.28%. As this takes place, the coefficient of chemical resistance was increased from 0.83 to 0.94, and after one year of exposure in the same corrosive environment, from 0.69 to 0.88.

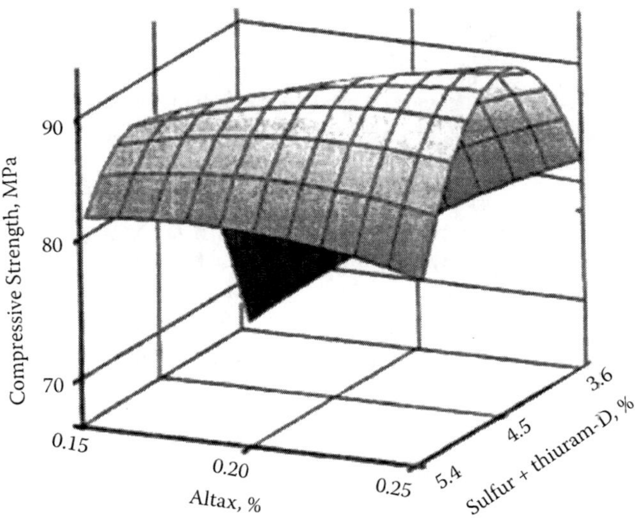

FIGURE 2.58 Compressive strength of RubCon samples relationship between quantities of Altax and sulfur + thiuram-D after exposition in 36% solution of a hydrochloric acid. (Reprinted from Yu. Potapov, Yu. Borisov, V. Chmyhv, and D. Beilin, "Research of Polymer Concrete Based on Low Molecular Polybutadiene, Part VIII: Chemical Resistance of Polymer Concrete," *J. Scientific Israel Technological Advantages* 7, nos. 1–2 (2005): 72–78. With permission.)

TABLE 2.10
Effect of Oxides and Salts of Multivalent Metals on Resistance of RubCon in 36% Solution of HCL

			Indexes		
			Strength, MPa`		
Test Number	Addition	Contains in the Mixture (%)	Initial	After 90 Days	K
1	TiO_2	1.5	103.6	93.2	0.9
		2.0	103.8	95.5	0.92
		2.5	104.4	97.1	0.93
		3.0	104.6	93.1	0.89
2	$BaSO_4$	2.0	101.9	89.7	0.88
		3.0	102.1	91.9	0.9
		4.0	101.8	93.7	0.92
		5.0	97.8	90.0	0.92
3	Control	103.0	83	0.83	

Source: Reprinted from V. Solomatov and V. Seljaev, "Chemical Strength of Composite Building Materials," *Stroiizdat*, Moscow, 1987 (in Russian). With permission.

Introduction of the other additives into the RubCon mix also raise its compressive strength. Our experiments have shown that addition of 2.3% TiO_2 or 3.6% $BaSO_4$ increases compressive strength of the RubCon samples after 90 days of exposure in a 36% solution of HCl up to 11%–12% (Table 2.10).

As illustrated in Figure 2.59 [1,21], an increase in compressive strength R of RubCon operated in a nonaggressive environment (line 1) is connected with an increase of the TiO_2 dose since metal oxides are activating agents of the vulcanization process. After 90 days of exposure to a 36% solution of HCl and a quantity of TiO_2 of more than 2.3%, the chemical resistance coefficient K_{cr} and compressive strength R decreased (lines 2, 3). We attribute this phenomenon to intensive formation of soluble compounds resulting in the production of microcracks.

The following equation was obtained from regression analysis:

$$R = -6.3x^2 + 28.4x + 64.5 \tag{2.63}$$

from which it follows that the optimal dose of TiO_2 is 2.3%.

In contrast to the TiO_2 additive, the introduction of $BaSO_4$ reduces the compressive strength R of RubCon operated in nonaggressive environments (Figure 2.60 line 1) because $BaSO_4$ acts as a fin-grained filler. With increasing doses of this additive, the discreteness of the rubber matrix grows, resulting in a reduction of RubCon strength. After 90 days of exposure to a 36% solution of HCl and a quantity of

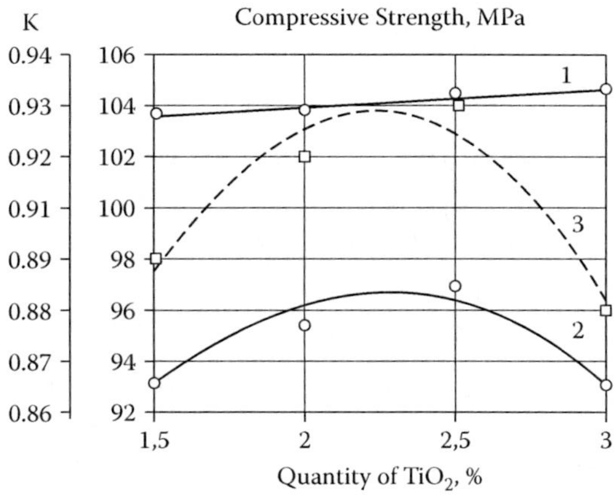

FIGURE 2.59 Impact of TiO_2 quantity on RubCon compressive strength operated in no aggressive environments (line 1), resistance coefficient (line 2), and compressive strength (line 3) after 90 days exposure to 36% solution of HCl. (Reprinted from Yu. Potapov, Yu. Borisov, V. Chmyhv, and D. Beilin, "Research of Polymer Concrete Based on Low Molecular Polybutadiene, Part VIII: Chemical Resistance of Polymer Concrete," *J. Scientific Israel Technological Advantages* 7, nos. 1–2 (2005): 72–78. With permission.)

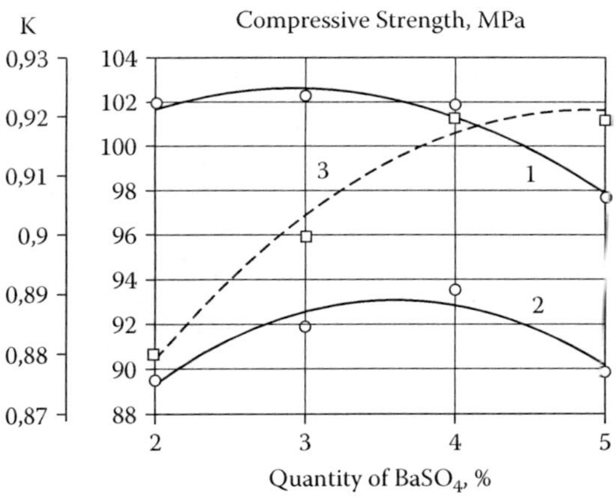

FIGURE 2.60 Impact of $BaSO_4$ quantity on RubCon compressive strength operated in no aggressive environments (line 1), compressive strength (line 2), and resistance coefficient (line 3) after 90 days exposure to 36% solution of HCl. (Reprinted from Yu. Potapov, Yu. Borisov, V. Chmyhv, and D. Beilin, "Research of Polymer Concrete Based on Low Molecular Polybutadiene, Part VIII: Chemical Resistance of Polymer Concrete," *J. Scientific Israel Technological Advantages* 7, nos. 1–2 (2005): 72–78. With permission.)

$BaSO_4$ of more than 3%, the chemical resistance coefficient K_{cr} of RubCon samples is increased (lines 3), whereas its compressive strength falls (line 2).

The regression model of these experiments is depicted by the equation

$$R = -1.5x^2 + 10.6x + 74.1, \tag{2.64}$$

from which it follows that the optimal dose of $BaSO_4$ is 3.6%.

Chemical Resistance of Steel Fiber–Reinforced RubCon

The corrosive environments that were used were water, a 70% solution of sulfuric acid, a 5% solution of phosphoric acid, a 36% solution of hydrochloric acid, and a 25% water solution of ammonia. Chemical resistance of fiber-reinforced RubCon was estimated on test specimens measuring 4 × 4 × 16 cm at exposures of 180 and 360 days at intermediate terms of 10, 30, 60, 90, 180, and 270 days.

Three samples of RubCon were made for each term of exposure. Before immersing in the medium, samples were measured and weighed. After reaction of the corrosion reagents with RubCon, the samples were taken out of the exicators, dried with filtering paper, and put on a compression test with a speed of 60 MPa/min.

Experimental curves of the mass transfer and chemical resistance of fiber-reinforced RubCon samples are illustrated in Figure 2.61.

The analysis of the obtained results shows that fiber-reinforced RubCon is a composite hydrophobic material with a coefficient of water resistance of K_{cr} = 0.995. Decreasing compressive strength was not observed and water absorption was 0.05% on weighing of samples. The small change of weight is due to the hydrophobic surface of RubCon. This is due to the intrinsic properties of the polybutadiene binder, which is not moistened with water. Furthermore, polybutadiene oligomer is nonpolar liquid.

A reduction of the physical–mechanical properties of RubCon samples in the first days of exposure is associated with penetration of water into the RubCon body

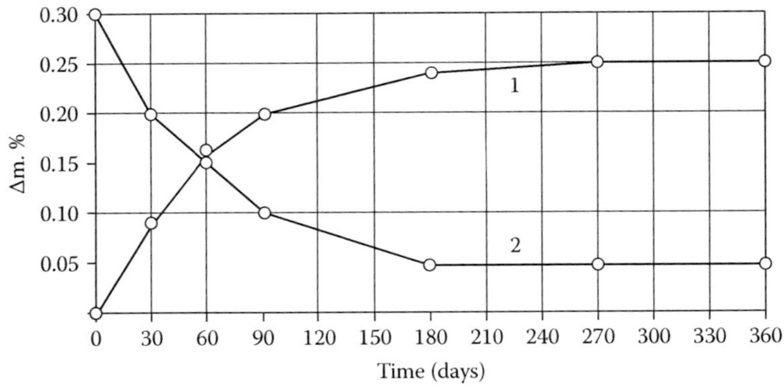

FIGURE 2.61 (1) Water absorbing of fiber reinforced RubCon, (2) coefficient of water resistance depending on exposure duration in water. (Reprinted from O. Figovsky and D. Beilin, "Building Materials Based on Advanced Polymer Matrix: Review," *J. Scientific Israel Technological Advantages* 10, no. 3 (2008): 1–119. With permission.)

through micropores and microcracks. As this takes place, mobility of structural elements and pressure in the tops of microcracks increase, resulting in a weakening of intermolecular bonds and decrease of strength.

Research shows that water has little or no effect on a decrease in strength and change of a material structure; in other words, the influence of water on the efficiency of RubCon structures is insignificant.

The destructive action of acids is caused by their nature, concentration, pH of water solutions, presence of oxidizing properties, temperature of the environment, and solubility of formed corrosion products at their interaction with polymer concretes and steel reinforcement.

Alkaline environments destroy the majority of composite building materials based on the organic matrix.

A water solution of ammonia was chosen as the aggressive media. Ions of ammonium NH_4^+ arises from dissolution of ammonia in water with the resulting formation of a hydroxide NH_4OH.

Results of the experiments are illustrated in Table 2.11 and Figures 2.62 and 2.63.

One can see that after 180 days of exposure of RubCon in aggressive environments, its compression strength decreased 22% in a 36% solution of hydrochloric acid, 8% in a 5% solution of a phosphoric acid, 7% in a 70% solution of a sulfuric acid, and 16% in a 25% solution of NH_4OH.

Destruction of the polymer binder is caused by inorganic acids that rupture polymer macromolecules and consequently increase deformability and decrease composite strength. Visual inspection of the samples after 180 days of exposure in inorganic acids show that steel fibers do not protrude, but the painted surface of the samples immersed in a 36% solution of hydrochloric acid changed significantly.

TABLE 2.11
Physical–Mechanical Properties of RubCon Samples at Action of Inorganic Acids and Solutions of the Alkalis

		Time of Exposure, Days					
Corrosive Media	**Indexes**	**0**	**10**	**30**	**50**	**90**	**180**
36% solution of HCl	$\Delta\, m$,%	0	−0.09	0.021	0.052	0.234	0.526
	K_{cr}	1	1.07	0.98	0.915	0.85	0.78
70% solution of H_2SO_4	$\Delta\, m$,%	0	−0.02	0.023	0.031	0.13	0.26
	K_{cr}	1	1.08	1.026	0.974	0.94	0.93
5% solution of H_3PO_4	$\Delta\, m$,%	0	−0.05	0.01	0.033	0.064	0.11
	K_{cr}	1	1.09	1	0.947	0.93	0.92
25% solution of NH_4OH	$\Delta\, m$,%	0	0.002	0.05	0.063	0.107	0.22
	K_{cr}	1	1.01	0.96	0.925	0.88	0.84

Source: Reprinted from O. Figovsky and D. Beilin, "Building Materials Based on Advanced Polymer Matrix: Review," *J. Scientific Israel Technological Advantages* 10, no. 3 (2008): 1–119. With permission.

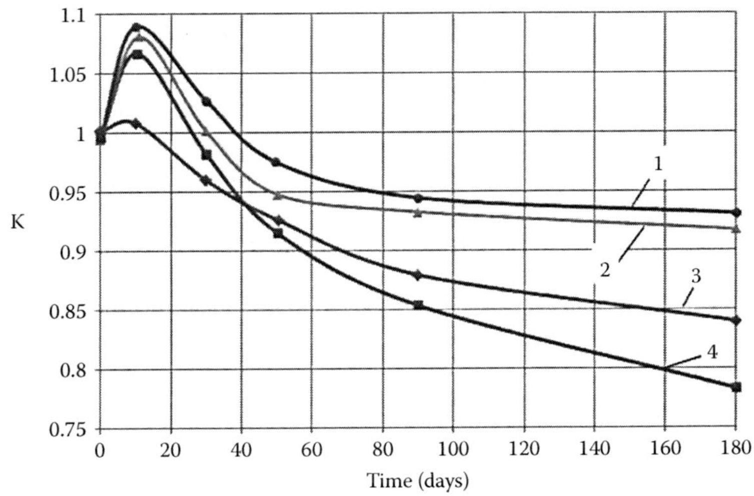

FIGURE 2.62 The coefficient chemical resistance dependence of the exposure time: (1) 70% solution of H_2SO_4, (2) 5% solution of H_3PO_4, (3) 25% solution of NH_4OH, (4) 36% solution of HCl. (Reprinted from O. Figovsky and D. Beilin, "Building Materials Based on Advanced Polymer Matrix: Review," *J. Scientific Israel Technological Advantages* 10, no. 3 (2008): 1–119. With permission.)

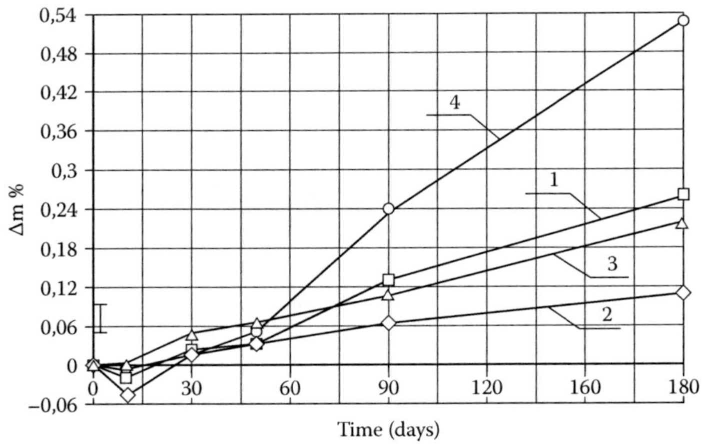

FIGURE 2.63 Kinetics of mass-absorption of RubCon: (1) 70% solution of a sulfuric acid, (2) 5% solution of a phosphoric acid, (3) 25% water solution of ammonia, (4) 36% solution of a hydrochloric acid. (Reprinted from O. Figovsky and D. Beilin, "Building Materials Based on Advanced Polymer Matrix: Review," *J. Scientific Israel Technological Advantages* 10, no. 3 (2008): 1–119. With permission.)

The analysis of obtained results show that the strength of the RubCon samples immersed in inorganic acids increased during the first 10 days (Table 2.11). For this period, compression strength increased by 7% in the 36% solution of a hydrochloric acid, by 9% in the 5% solution of a phosphoric acid, and by 8% in the 70% solution of a sulfuric acid.

Increase of the RubCon sample strength in an initial stage of interaction with the aggressive liquid environments is due to absorption of the liquid by structural microdefects that are responsible for superficial compacting. During long-term exposure to aggressive environment oxidizers, such as mineral acids (hydrochloric, phosphoric concentrated sulfuric), alkalis, and so on, deep structural changes occur in composites and worsen their operational properties. Destruction of the polymeric binder and rupture of adhesive bonds cause a decrease in polymer strength.

The action of a *sulfuric acid* on RubCon is accompanied by an insignificant reduction mass of the samples in the first days of exposure; however, in the process of dissolution of steel fibers, the mass absorption process becomes obvious. As this takes place, hydrogen sulfide and sulfurous anhydride are released and can create possible sulfiding.

A firm thermoplastic mass, identical in structure to initial rubber but with a smaller content of double bonds, is developed under the influence of the concentrated sulfuric acid. A decrease in the quantity of double bonds in a rubber molecule structure leads to decreased strength parameters of vulcanizates.

The *hydrochloric acid*, dry and damp chlorine, causes fast destruction of most rubbers regardless of the type of filler. Chemical resistance of ebonites (i.e., highly filled composites) on the rubber base (including RubCon) to hydrochloric acid is rather high, which permits their application for corrosion protection of building structures and materials. It is worth noting that at an early stage of interaction of the acid with RubCon samples, a process of mass absorption damps. In this case, processes of oxidation and isomerization are observed; products of a chlorination layer are produced on the ebonite surface, which inhibits penetration of the material by the aggressive media. Therefore, the service life of the composite depends on the speed of chlorination and oxidation processes and the density of the powders formed on its surface.

The mass absorption process of RubCon samples immersed in *phosphoric acid* goes more slowly because of the low penetration ability of the acid.

Thus, decreases in the physical–mechanical characteristics of RubCon under the action of inorganic acids is caused by various speeds of chemical reactions such as corrosion, oxidation, sulfiding, chlorination, and so on. Processes of oxidation and depolymerization proceed with formation of low-molecular compounds containing hydroxyl and carboxyl groups. The process of oxidation has a branched-out character.

Intensive destruction of the rubber-based polymer by *alkalis* occurs due to formation of an unstable complex that breaks up with regeneration of initial substances and following insignificant isomerization. It is believed that the nitrogen contained in a 25% water solution of ammonia will form nitrocompounds, which at interaction with vulcanizates, initiate nitriding with the resulting destruction of a polymeric chain.

Effect of γ-Radiation on the Structure and Properties of RubCon γ-Radiation Shielding

Rapid growth of nuclear power has brought to a head the problem of storage and burial of radioactive waste products. Now, more than 250,000 tons of exhausted nuclear fuel has accumulated in the world. In light of this, creation of new radioactive-resistant structural materials intended for protection, storage, and burial of radioactive waste products is necessary. Application of RubCon for protection against radiation in enclosed radioactive waste product storehouses is discussed below.

The effect of γ-radiation on the RubCon structure was studied by infrared spectroscopy in an area 4,000÷400 cm^{-1}. Spectrums of the RubCon samples demonstrate the essential changes of the material structure after γ-radiation exposure (Figure 2.64) [25,26].

The radiation protective properties of RubCon with regard to γ-radiation are defined by linear attenuation coefficient (LAC) value. LAC is the actual fraction of photons interacting per unit thickness of material. Linear attenuation coefficient values indicate the rate at which photons interact as they move through material. LAC depends on properties of the material and radiant energy.

It is known that

$$\mu = n\sigma. \tag{2.65}$$

where n is the number of atoms of absorbing medium in units by volume, and σ is the cross-section area of radiation interaction with one atom of absorber. Here, σ depends on the atomic number of the material and radiant energy. Experimental values of LAC for RubCon were obtained in a range of radiant energy of 0.1–1.5 MeV

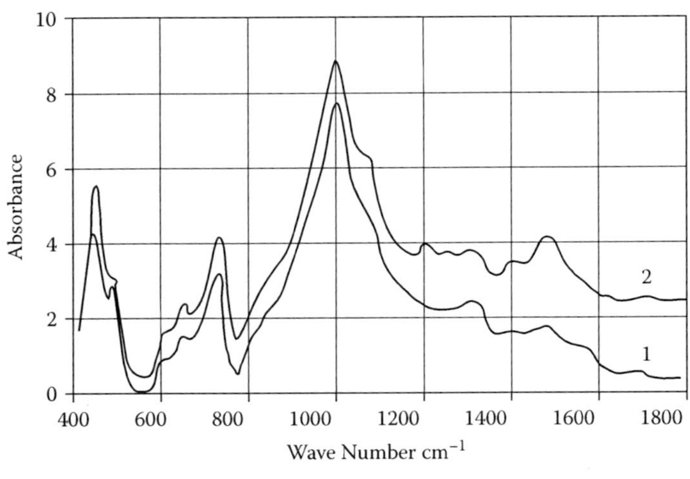

FIGURE 2.64 Specters of RubCon: (1) before γ-exposition, (2) after γ-exposition. (Reprinted from O. Figovsky and D. Beilin, "Building Materials Based on Advanced Polymer Matrix: Review," *J. Scientific Israel Technological Advantages* 10, no. 3 (2008): 1–119. With permission.)

of basic radioactive isotopes, which determine radiation in storehouses of radioactive waste products.

Attenuation of monoenergetic gamma rays collimated into a narrow beam provided the basis for experimental investigation of LAC. The process of ray attenuation in their passage through the shield is described by equation (2.66) [28]:

$$I_x = I_0 \, e^{-\mu x} \tag{2.66}$$

where I_x is the photon intensity outside of a shield, I_0 is the unshielded photon intensity, μ is the linear attenuation coefficient, and x is the thickness of the shielding material. This decreasing exponential absorption characteristic is shown in Figure 2.65.

From (2.60) it follows that

$$\mu = (\ln I_0 - \ln I_x)/x \tag{2.67}$$

The linear attenuation coefficient is the sum of the probabilities of interaction per unit path length by each of the three scattering and absorption processes: photoelectric effect, Compton effect, and electron–positron pair production. The reciprocal of μ is defined as the mean-free path, which is the average distance the photon travels in an absorber before an interaction takes place.

Experimental determination of LAC was carried out using the scintillation spectrometer of gamma radiation at power 1.2 MeV (Figures 2.66 and 2.67).

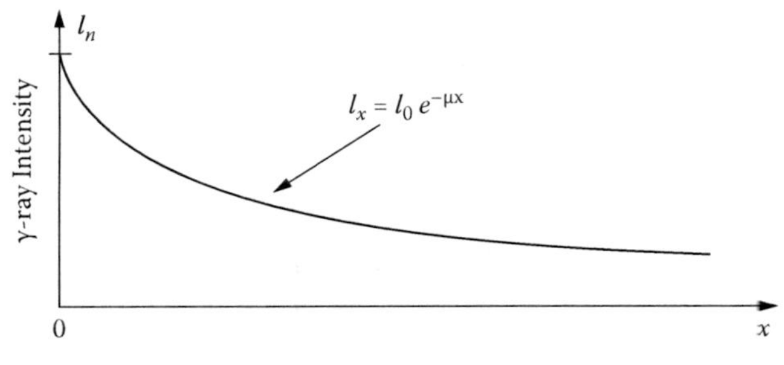

FIGURE 2.65 The gamma rays intensity dependence of the thickness of shielding material. (Reprinted from Yu. Potapov, Yu. Borisov, V. Chmyhv, and D. Beilin, "Research of Polymer Concrete Based on Low Molecular Polybutadiene, Part VIII: Chemical Resistance of Polymer Concrete," *J. Scientific Israel Technological Advantages* 7, nos. 1–2 (2005): 72–78. With permission.)

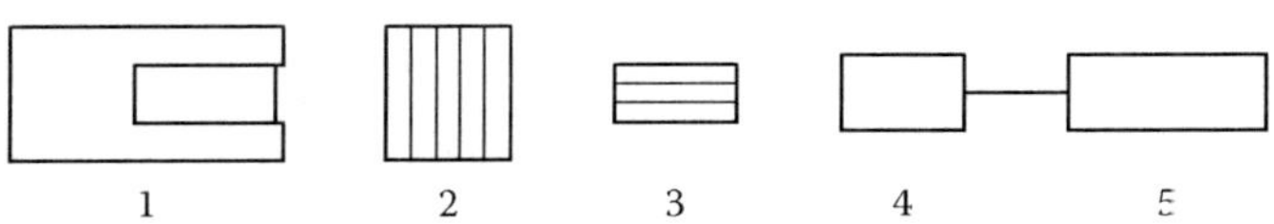

FIGURE 2.66 Block diagram of LAC meter: (1) Source and collimator of γ -radiation, (2) absorber, (3) collimator, (4) scintillator, (5) analyzer. (Reprinted from Yu. Potapov, Yu. Borisov, V. Chmyhv, and D. Beilin, "Research of Polymer Concrete Based on Low Molecular Polybutadiene, Part VIII: Chemical Resistance of Polymer Concrete," *J. Scientific Israel Technological Advantages* 7, nos. 1–2 (2005): 72–78. With permission.)

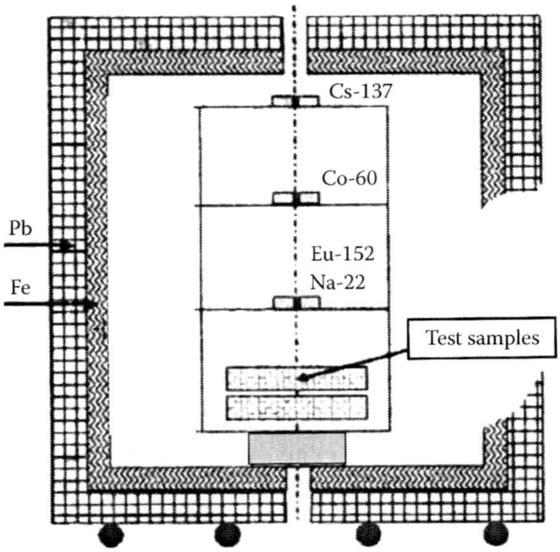

FIGURE 2.67 Plant for measurement of LAC. (Reprinted from O. Figovsky and D. Beilin, "Building Materials Based on Advanced Polymer Matrix: Review," *J. Scientific Israel Technological Advantages* 10, no. 3 (2008): 1–119. With permission.)

The experimental values of RubCon LAC make it possible to estimate its γ-radiation absorption from the formula

$$K = 1 - e^{-\mu x}, \tag{2.68}$$

where K is the γ-radiation absorption.

Results of the experiment are illustrated in Figures 2.68 and 2.69 and Table 2.12.

For a RubCon sample thickness of $x = 1$ cm, results are shown in Table 2.13. Table 2.14 illustrates the absorbing properties of investigated materials with cement concrete (for a narrow radiation beam). The material layers necessary for attenuation of the primary flux of γ-quantums in 2, 10, and 100 times are shown.

One can see that the γ-radiation absorption of RubCon is close to that of ordinary Portland cement concrete. It is known that LAC depends on the density of the shielding material. Consequently, increasing γ-radiation absorption of RubCon is accomplished by the appropriate selection of filler, coarse and fine aggregates.

Radioactive Resistance of RubCon

The radioactive resistance of a structural material is its ability to maintain its physical–mechanical and operational properties within given limits in time and after an ionizing irradiation.

The estimation of radioactive resistance of RubCon was done by studying its compressive strength and modulus of elasticity depending on the absorbed dose of radiation. The source of radiation was the isotope Co^{60}. The maximum absorbed

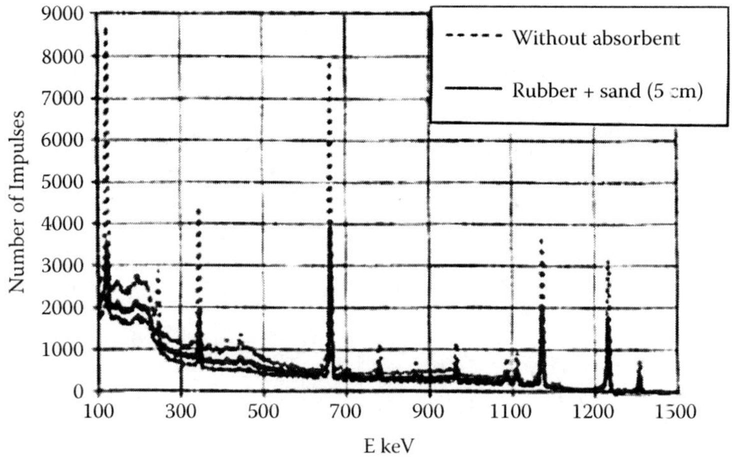

FIGURE 2.68 Specters of γ-radiation passing through rubber with sand layers. (Reprinted from O. Figovsky and D. Beilin, "Building Materials Based on Advanced Polymer Matrix: Review," *J. Scientific Israel Technological Advantages* 10, no. 3 (2008): 1–119. With permission.)

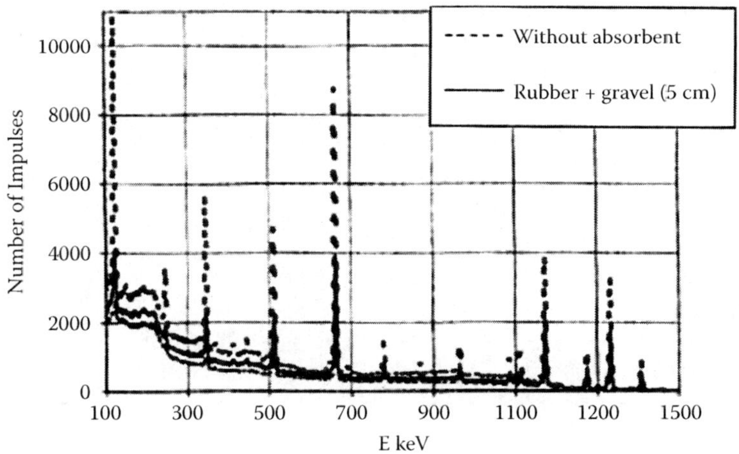

FIGURE 2.69 Specters of γ-radiation passing through rubber with gravel layers. (Reprinted from O. Figovsky and D. Beilin, "Building Materials Based on Advanced Polymer Matrix: Review," *J. Scientific Israel Technological Advantages* 10, no. 3 (2008): 1–119. With permission.)

doze was $50*10^6$ R. RubCon samples measuring 40 x 40 x 160 mm were tested for axial compression at a constant speed of loading of 60 MPa/min of loading exposure at each stage of the test range up to 30–40 seconds. It was noted that the temperature of samples due to radiation heating did not exceed 50°C during the experiment.

The experiments show that the coefficient of radioactive resistance of RubCon at an absorbed dose of $50*10^6$ *R* is equal to 1.0.

TABLE 2.12
Experimental Values of LAC of RubCon and Some Other Materials

Power of γ -radiation (MeV)	Isotope	μ (cm^{-1})		
		Aluminum	Portland Cement Concrete, g = 2.3 t/m³	RubCon
0.122	Eu^{152}	0.417	0.355	**0.26**±0.02
0.662	Cs^{137}	0.207	0.177	**0.14**±0.01
1.173	Co^{60}	0.156	0.132	**0.12**±0.01
1.332	Co^{60}	0.147	0.124	**0.11**±0.01

Source: Reprinted from Yu. Potapov, Yu. Borisov, V. Chmyhv, and D. Beilin. "Research of Polymer Concrete Based on Low Molecular Polybutadiene, Part VIII: Chemical Resistance of Polymer Concrete," *J. Scientific Israel Technological Advantages* 7, nos. 1–2 (2005): 72–78. With permission.

TABLE 2.13
Absorption of Radiation of RubCon and Some Other Materials

Power of γ -radiation (MeV)	*K*		
	Aluminum	Portland Cement Concrete, g = 2.3 t/m³	RubCon
0.122	0.341	0.299	0.23
0.662	0.188	0.162	0.13
1.173	0.144	0.124	0.11
1.332	0.136	0.117	0.10

Source: Reprinted from Yu. Potapov, Yu. Borisov, V. Chmyhv, and D. Beilin. "Research of Polymer Concrete Based on Low Molecular Polybutadiene, Part VIII: Chemical Resistance of Polymer Concrete," *J. Scientific Israel Technological Advantages* 7, nos. 1–2 (2005): 72–78. With permission.

Effect of γ-Radiation on Mechanical and Chemical Properties of RubCon

The influence of γ-radiation on stress–strain properties of RubCon is plotted in Figures 2.70 and 2.71.

The analysis of the results shows that values of relative longitudinal and lateral strains and the modulus of elasticity of RubCon samples at compression before and after irradiation are essentially constant. At significant loading, corresponding to a compressive strength of 90 MPa, relative longitudinal strains of irradiated samples (absorbed dose $50{*}10^6$ *R*) is about 10% lower, and the value of the modulus of elasticity is higher by 7% in comparison with the same parameters of the control samples before irradiation. It may be concluded that hardening of RubCon is a result of structural changes under the influence of radiation.

TABLE 2.14
Comparison of the Absorbing Properties of RubCon and Portland Cement Concrete

$E\gamma$, (MeV)	μ (cm^{-1})	$x_{1/2}$	x_{10} (cm)	x_{100}
Cement Concrete, g = 2.3 g/cm³				
0.1	0.378	1.834	6.091	12.180
0.2	0.275	2.520	8.373	16.750
0.5	0.194	3.573	11.870	23.740
1.0	0.141	4.916	16.330	32.660
1.5	0.116	5.975	19.850	39.700
RubCon with Sand, g = 2.0 g/cm³				
0.1	0.351	1.976	6.563	13.126
0.2	0.255	2.719	9.032	18.063
0.5	0.177	3.924	13.035	26.070
1.0	0,129	5.389	17.901	35.801
1.5	0.105	6.617	21.981	43.961
RubCon with Gravel, g = 2.1 g/cm³				
0.1	0.398	1.743	5.789	11.577
0.2	0.273	2.541	8.442	16.885
0.5	0.187	3.711	12.327	24.655
1.0	0.136	5.102	16.950	33.900
1.5	0.110	6.281	20.865	41.731

Source: Reprinted from O. Figovsky and D. Beilin, "Building Materials Based on Advanced Polymer Matrix: Review," *J. Scientific Israel Technological Advantages* 10, no. 3 (2008): 1–119. With permission.

Note: $x_{1/2}$ – layer for attenuation in ½ times; x_{10} – layer for attenuation in 10 times; x_{100} – layer for attenuation in 100½ times.

The effect γ-radiation on *compressive strength* of RubCon is shown in Figure 2.72. It is worth noting that all samples kept their color and geometric parameters after exposure. This can be seen graphically in Figure 2.71 that the compressive strength of RubCon samples decreased no more than 12% less than their initial strength.

The established change of compression strength depending on the absorbed dose is adequately described by this mathematical model:

$$y = -0.09x^4 + 1.04x^3 - 3.06x^2 + 1.59x + 88.43 \tag{2.69}$$

where y is the compressive strength of RubCon and x is the absorbed dose.

It is known that a material is considered as radiation resistant if its strength parameters decrease no more than 25% from the initial value. Thus RubCon can be classified as a material that is resistant to the influence of γ-radiation at an absorbed

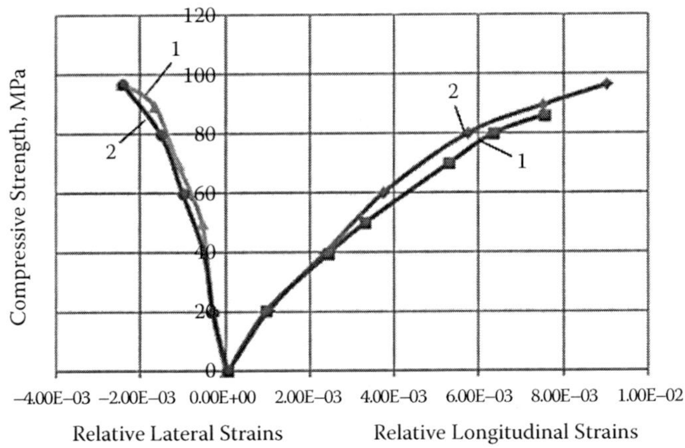

FIGURE 2.70 Relative longitudinal and lateral strains of RubCon: (1) before irradiation, (2) after irradiation. (Reprinted from Yu. Potapov, Yu. Borisov, V. Chmyhv, and D. Beilin, "Research of Polymer Concrete Based on Low Molecular Polybutadiene, Part VIII: Chemical Resistance of Polymer Concrete," *J. Scientific Israel Technological Advantages* 7, nos. 1–2 (2005): 72–78. With permission.)

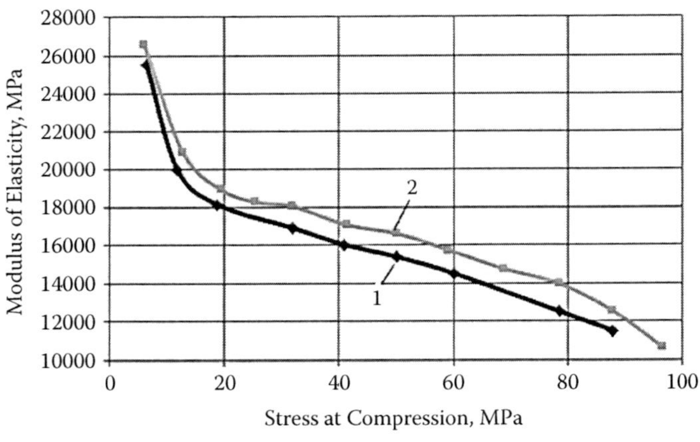

FIGURE 2.71 Modulus of elasticity of RubCon: (1) before irradiation, (2) after irradiation. (Reprinted from Yu. Potapov, Yu. Borisov, V. Chmyhv, and D. Beilin, "Research of Polymer Concrete Based on Low Molecular Polybutadiene, Part VIII: Chemical Resistance of Polymer Concrete," *J. Scientific Israel Technological Advantages* 7, nos. 1–2 (2005): 72–78. With permission.)

dose up to 8.4 MGy, and therefore may be used as a structural material in construction of radioactive waste storehouses.

The combination of *chemically aggressive environments* and ionizing radiation was experimentally studied. The absorbed dose of γ-radiation was 5 Gy; the prevalence of the aggressive environment was its selection criterion. The results of experimental investigation are demonstrated in Table 2.15.

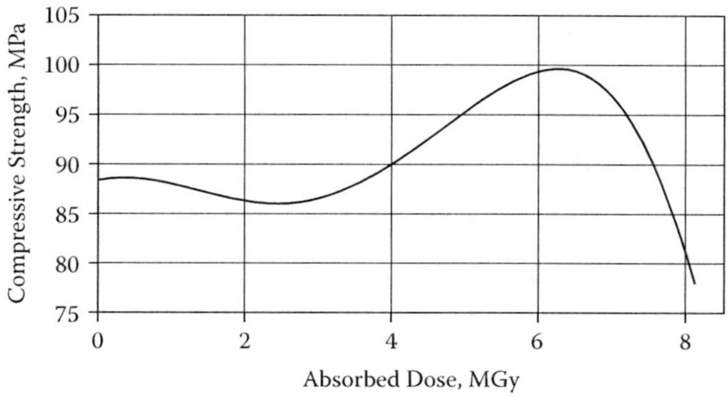

FIGURE 2.72 Effect γ-radiation on compressive strength of RubCon. (Reprinted from O. Figovsky and D. Beilin, "Building Materials Based on Advanced Polymer Matrix: Review," *J. Scientific Israel Technological Advantages* 10, no. 3 (2008): 1–119. With permission.)

TABLE 2.15
Effect of γ-Radiation on the Coefficient of Chemical Resistance of RubCon

Aggressive Environment (1-year exposition)	Coefficient of Chemical Resistance: γ- Radiation Before (K_0)	After (K_1)	($K_0 - K_1$)100%
70% solution of a sulfuric acid	0.920	0.920	0
5% solution of a phosphoric acid	0.935	0.900	−3.7
3% solution of nitric acid	0.810	0.800	−1.2
36% solution of a hydrochloric acid	0.690	0.690	0
5% solution of an acetic acid	0.816	0.780	−4.4
10% solution of a lactic acid	0.950	0.940	−1
10% solution of a citric acid	0.873	0.870	−0.3
25% water solution of ammonia	0.815	0.790	−3.1
10% solution of caustic sodium	0.871	0.820	−5
10% solution of potassium hydroxide	0.910	0.900	−1.1
Saturated water solution of sodium chloride	0.957	0.950	−0.7
30% solution of copper sulfate	0.835	0.816	−2/3
Acetone	0.881	0.830	−5.8
Diesel fuel	0.878	0.850	−3.2

Source: Reprinted from O. Figovsky and D. Beilin, "Building Materials Based on Advanced Polymer Matrix: Review," *J. Scientific Israel Technological Advantages* 10, no. 3 (2008): 1–119. With permission.

TABLE 2.16
Componential Structure of RubCon with Lead Pellet

Component	Quantity
Polybutadene rubber	4.86
Sulfur technical	2.05
Thiuram-D	0.28
Zinc salt of 2- mercaptobenzthiazole	0.23
Zinc oxide	0.80
Calcium oxide	0.15
Fly ash	4.32
Quartz sand	13.12
Lead pellet	74.01

Source: Reprinted from O. Figovsky and D. Beilin, "Building Materials Based on Advanced Polymer Matrix: Review," *J. Scientific Israel Technological Advantages* 10, no. 3 (2008): 1–119. With permission.

The analysis of experimental data shows (Table 2.15) that RubCon displays universal chemical stability and keeps it after γ-radiation exposure. Coefficients of chemical resistance of RubCon were above 0.80 in all environments except for a 36% solution of hydrochloric acid, where it was equal to 0.69.

Further research has shown that replacement of granite rubble by quartz sand allows an increase if the coefficient of chemical resistance of RubCon up to 7% in a 36% solution of hydrochloric acid.

Increase of a material density by introduction of a heavy filler is the obvious way to raise its radiation resistance properties. The new component structure of RubCon having the highest chemical and radiation resistance is shown in Table 2.16.

The main physical–mechanical characteristics of the new composite are shown in Table 2.17.

Based on research, it may be inferred that RubCon with lead-containing filler has high radiation protective characteristics and practically universal chemical stability.

Summary

- RubCon offers universal chemical resistance many times higher than ordinary Portland cement concrete and surpasses the chemical resistance of polyester, polyepoxy, and vinyl ester polymer concretes.
- The most intensive decrease of RubCon strength occurs during the first 6 months of exposure to a corrosive environment.

TABLE 2.17
The Main Physical–Mechanical Characteristics of the New RubCon Structure

Properties	Value
Average density, kg/m^3	4.86
Compression strength, MPa	2.05
Bending strength, MPa	0.28
Water absorption	0.23
Maximal operation temperature,°C	0.80
Linear attenuation coefficient, cm^{-1}	0.15
Coefficient of radiation resistance at γ-radiation dose, MGy	4.32

Source: Reprinted from O. Figovsky and D. Beilin, "Building Materials Based on Advanced Polymer Matrix: Review," *J. Scientific Israel Technological Advantages* 10, no. 3 (2008): 1–119. With permission.

- Elevated temperature (up to 60°C) insignificantly reduces the chemical resistance of RubCon samples and leads to their destruction due to acceleration of chemical reactions between the corrosive environment and the material.
- The chemical resistance coefficient of RubCon operating in aggressive environments over 10 years exceeds 0.5. Obtained data allow determination of the thickness of the concrete cover of a reinforced RubCon structure.
- γ-radiation absorption of RubCon is close to that of ordinary Portland cement concrete.
- Increasing of γ-radiation absorption of RubCon is rendered by the appropriate selection of filler, coarse and fine aggregates.
- The temperature of RubCon samples did not exceed 50°C due to radiation heating at an absorbed dose of $50*10^6$ R.
- The coefficient of radioactive resistance of RubCon at an absorbed dose of $50*10^6$ R is equal to 1.0.

MANUFACTURING PROCESS OF RUBCON STRUCTURES AND PRODUCTS [1–3,24,31]

Technological Manufacturing of RubCon

Preparation of the initial materials includes

- Drying and sieving of fillers by fractions
- Drying of the aggregates and components of the curing groups
- Storage and consumption of each kind of component

Preparation of RubCon compositions in view of polystructure principles assumes application of separate technology.

According to this technology the binder is prepared in a high-speed mixer. The measured portions of liquid rubber, curing group components, and filler are fed into the mixer from a warehouse. After that, the received binder and aggregate are fed into the usual mixer. The aggregate is loaded directly after drying at a temperature of 70±5°C to decrease the mixture viscosity, consumption of the rubber, and for improvement of the mixture formation and packing. The mixture then moves in warmed metal forms at a temperature of 60±5°C.

Silicon liquids (a solution of silicon-organic rubber in gasoline) and polyethylene terephthalate film are used to prevent sticking of finished articles to forms. It is worth noting that polyester film is most effective. With its application, RubCon articles are easily removed from the form and have a brilliant surface.

The molded articles are placed in a chamber for thermal processing at temperatures of 100°C–130°C for a period of 2–18 hours depending on the thickness and weight of a product. After thermal processing, the forms are disassembled and directed to a preparation area.

A diagram of the technological process of RubCon manufacture is shown in Figure 2.73.

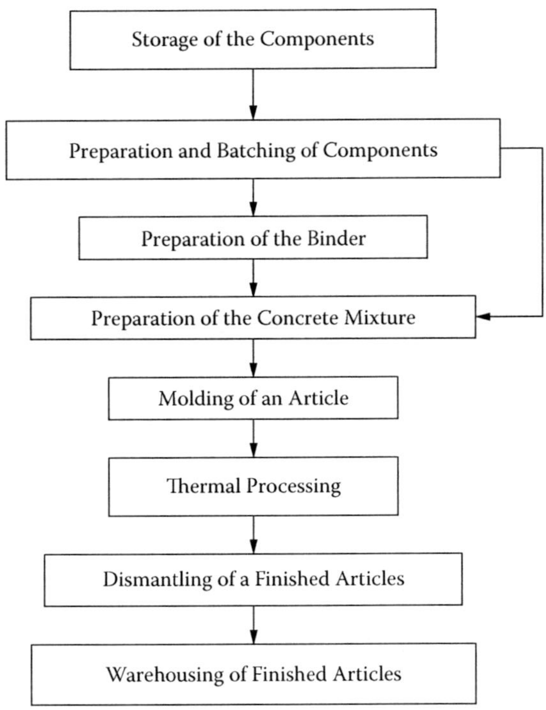

FIGURE 2.73 Technological process of RubCon manufacturing. (Reprinted from O. Figovsky and D. Beilin, "Building Materials Based on Advanced Polymer Matrix: Review," *J. Scientific Israel Technological Advantages* 10, no. 3 (2008): 1–119. With permission.)

RubCon Mixture Agitation Process

The dominating factor of effective composition preparation is the mode of mixing. During mixing of granular and fibrous components with liquid rubber, the critical stresses arise in the latter rubber. These stresses negatively influence the mixture. With increasing speed, the rotation of the mixer's rotor (its speed of deformation) is accompanied by increasing shear stresses, relaxation of the polymer does not have time to develop, and as a result, the polymer's flow properties. In these conditions, the mixture is pulled away from the mixer walls and agitation does not occur.

Usually, critical stresses corresponding with critical speeds of shear deformations fall within a range of 0.1–0.3 MПa and do not depend on molecular weight and molecular mass distribution of the polymer; however, these critical stresses decrease during introduction of solvent (plasticizer) in the system. On the contrary, critical deformation speeds are very sensitive to the molecular structure of the polymer (viscosity). This effect opens the practice opportunities of directed regulation of the critical deformation conditions. A combination of critical stresses and raised temperature can cause the mixture to begin to set before it reaches sufficient homogeneity. The greatest homogeneity can be obtained only at certain mixing parameters.

The lowest power expense and maximum homogeneity of the mix can be attained by optimum agitation of the rubber binder, fillers, and reinforcing fibers with the Brabender plastograph mixer. Such mixers allow estimation of the technological properties of rubber compositions based on viscous resistance, which in turn depends on duration of mechanical effects and temperatures.

The typical curves corresponding to the critical and optimum modes of mixing are presented on Figure 2.74.

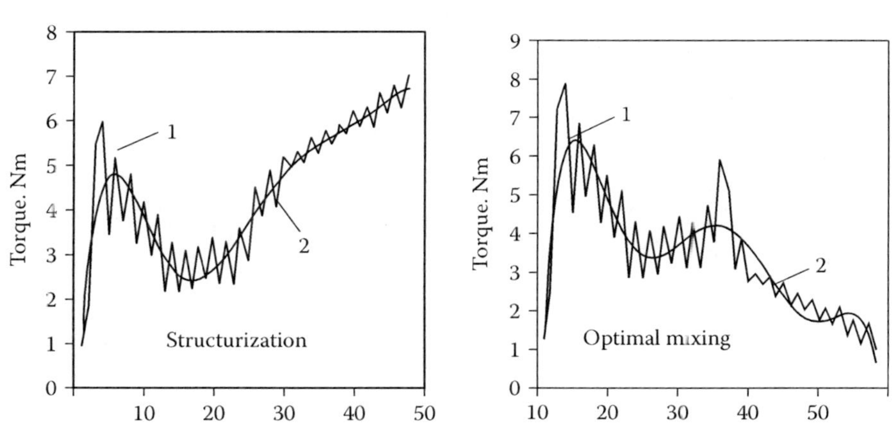

FIGURE 2.74 Change of the torque depending of mixingparameters: (1) automatic recorder, (2) approximating curve. (Reprinted from O. Figovsky and D. Beilin, "Building Materials Based on Advanced Polymer Matrix: Review," *J. Scientific Israel Technological Advantages* 10, no. 3 (2008): 1–119. With permission.)

The cumulative analysis of experimental data allows estimation of the quality of prepared compositions and various mixing duration and intensity. The experimental data on modeling dependences make it possible to optimize the specified processes.

The superficial phenomena of physical-chemical interaction of the liquid and solid phases and formation of the composite material microstructures proceed most intensively during rubber binder preparation. Therefore, the conditions under which the binder is prepared exert essential influence on the properties of composites. Uniformity of the prepared binder is defined by the intensity of the agitation process and depends on the process's speed and duration.

The object of research was a mix of low-molecular oligodiene, curing groups, and fly ash ablation with a specific surface of approximately 300 m^2/kg. Preparation of the binder was carried out in the laboratory rotary mixer. Times of agitation were 60, 180, and 300 s, at speeds of 8.3, 16.6, and 25 rps, respectively.

RubCon samples measuring 4 × 4 × 16 cm were molded from the resulting mixtures. After curing, the samples were weighed and tested for compression and tension at bend.

Curves based on power consumption versus time and speed of agitation are shown in Figure 2.75. The values of power input related to the volume of a mix (specific power input) as a function of speed and time of agitation are presented in Table 2.18.

The coefficients of mixture homogeneity, compressive strength, and average density of composite versus specific power input (SPI) are presented in Figure 2.76.

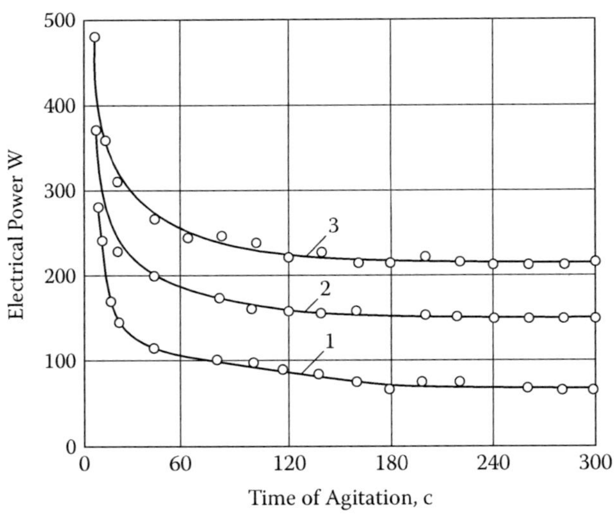

FIGURE 2.75 Dependence of power consumption on time and speed of binder agitation: (1)–(3) Speed of agitation: 8.3, 16.6, 23.3 rps, respectively. (Reprinted from O. Figovsky and D. Beilin, "Building Materials Based on Advanced Polymer Matrix: Review," *J. Scientific Israel Technological Advantages* 10, no. 3 (2008): 1–119. With permission.)

TABLE 2.18
Specific Power Input

Time of Agitation,(s)			
300	**180**	**60**	**Speed of Agitation, (rps)**
5.16	3.66	1.92	8.3
9.90	6.38	2.71	16.6
13.70	8.80	3.06	25.0

Source: Reprinted from O. Figovsky and D. Beilin, "Building Materials Based on Advanced Polymer Matrix: Review," *J. Scientific Israel Technological Advantages* 10, no. 3 (2008): 1–119. With permission.

It can be seen in Figure 2.76 that the coefficient of mixture homogeneity is stabilized at a level of 5.5% with SPIs of 3.6–4 kW*h/m^3, and at additional power inputs there is essentially no change. Changes in the compressive strength parameters with additional SPIs are a result of balancing three basic processes that begin at agitation:

- Movements, repacking, and regrouping of solid particles in a liquid phase
- Actions of the superficial interphase phenomena of wetting, spreading, and adhesions on a border of oligomer filler
- Air entrainment and decreasing the average density of the binder

The first two processes positively influence the structure formation of a composite material; the third produces a destructive effect. At SPI = 1.92–2.71 кW*h/m^3, the strength of the composite increases slightly due to improvement in mixture uniformity. Thus, the process of air entrainment begins and the density of RubCon decreases.

The uniformity and density of the mixture do not change significantly when the power inputs are increased from 3.06 to 3.66 кWh/м^3. However, compressive strength increases sharply and reaches the maximum value at SPI = 3.66–3.8кWh/м^3, due intensification of wetting of filler particles by oligomer. The structure of the oligomer boundary layers is transformed into an oriented condition by the action of energetic fields of filler particles. A new phase of the polymeric matrix is formed. At SPI = 3.9–6.38 кWh/м^3, strength parameters are decreased the density of the composite is reduced due to significant air entrainment.

The generalized dependence of specific power inputs on time of mix agitation is plotted in Figure 2.77.

Knowing the optimal value of specific power inputs makes it possible to define the modes of liquid rubber binder preparation.

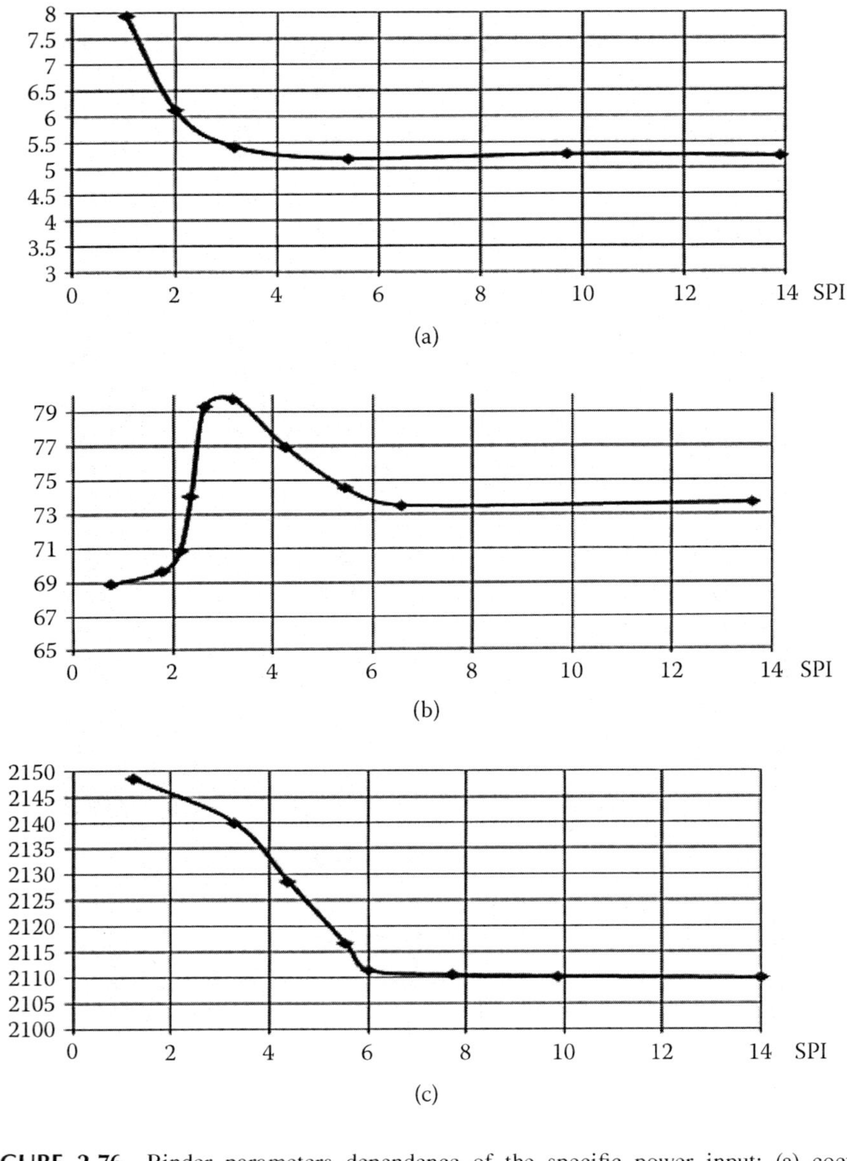

FIGURE 2.76 Binder parameters dependence of the specific power input: (a) coefficient of mixture homogeneity, (b) compressive strength, (c) average density of composites. (Reprinted from O. Figovsky and D. Beilin, "Building Materials Based on Advanced Polymer Matrix: Review," *J. Scientific Israel Technological Advantages* 10, no. 3 (2008): 1–119. With permission.)

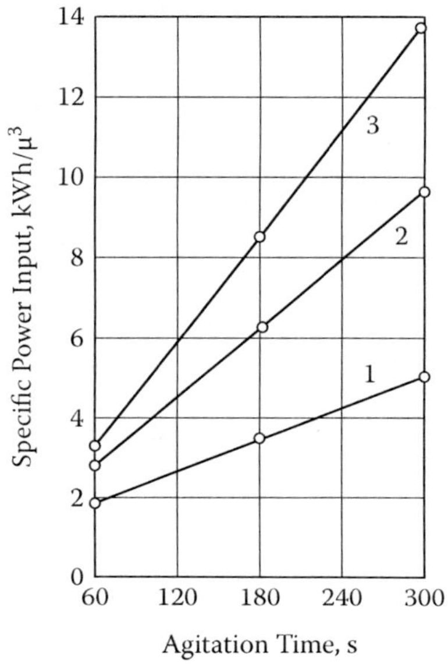

FIGURE 2.77 Dependence of specific power input on time of mix agitation for different rotational speeds: (1) 8.3 rps, (2) 16.6 rps, (3) 23.3 rps. (Reprinted from O. Figovsky and D. Beilin, "Building Materials Based on Advanced Polymer Matrix: Review," *J. Scientific Israel Technological Advantages* 10, no. 3 (2008): 1–119. With permission.)

It was determined that a uniform mix and composite with maximum strength parameters can be produced at SPI = 3.6–3.8 кWh/м³.

Influence of Temperature on RubCon Binder Hardening and Ways to Reduce It

The manufacturing finishing process of rubber concrete is vulcanization of the mixture. The mode of hardening (vulcanization) and RubCon's final characteristics depend on the temperature and heating time of RubCon articles. The structure of the composite depends on the time delay before the mixture is heated, the environment in which it is hardened, and the temperature and pressure exerted during hardening.

Reduction in the temperature of vulcanization is the important practical problem, and decisions on hardening temperatures allows simplification of vulcanization technology. At temperatures below a certain limit, it is not possible to reach the necessary degree of hardening and to obtain three-dimensional polymers with the required properties. Higher temperatures reduce processing time, but the faster hardening makes a course of relaxation processes difficult and leads to formation of strained (defective) structures and the smooth hardening may result in rigid composite structures.

TABLE 2.19
Results of Two-Factor Experimental Plan

Strength, MPa		Varied Parameters	
Tension at Bend	**Compression**	**Time, Hours**	**Temperature °C**
25.14	61.0	5	130
32.14	79.8	9	90
25.00	71.49	9	130
28.35	20.47	5	90
28.54	85.53	7	130
37.67	64.33	7	90
28.78	83.66	9	110
30.03	82.71	5	110
31.14	86.66	7	110

The temperature and time of vulcanization of RubCon samples were accepted as the basic variable factors; optimization parameters were compression strength and tension strength at bend of the samples.

The results of experimental research are shown in Table 2.19.

The compression strength σ_c dependence on vulcanization temperature mode according to experimental data is:

$$\sigma_c = -999.3 + 13.72\,X_1 + 83.14\,X_1 - 0.05\,X_1^2 - 2{,}96\,X_2^2 - 0.33\,X_1X_2 \quad (2.70)$$

where X_1 and X_2 are temperature and vulcanization time, respectively.

The response surface of vulcanization mode is illustrated in Figure 2.78.

Analysis of the data shows that the maximum compression strength is reached at a vulcanization temperature in the area of 110°C over 7–7.5 hours.

The following optimized parameter was tension strength at a bend σ_{tb}. The equation of regression is

$$\sigma_{tb} = -212.2 + 1.95\,X_1 - 1.37\,X_2 - 0.004\,X_1^2 - 1.34\,X_2^2 - 0.15\,X_1X_2 \quad (2.71)$$

where X_1 and X_2 are temperature and vulcanization time, respectively.

It is easy to verify that the maximum tension strength at a bend is reached at a vulcanization temperature of 90°C over 8 hours, at a temperature of 110°C over 7 hours, and at a temperature of 130°C over 6 hours.

Thus, the area of optimum vulcanization temperatures is 90°C–115°C and the optimum duration of heat treatment is 6–8 hours (Figure 2.79).

In light of this, a search for ways of reducing vulcanization temperature is gaining importance. Research shows that lowering the curing temperature of a RubCon mixture can be achieved in two ways:

1. Designing the curing group structure such that the temperature of mixture vulcanization is minimal.
2. Regulation of a microstructure and viscosity of the liquid rubber.

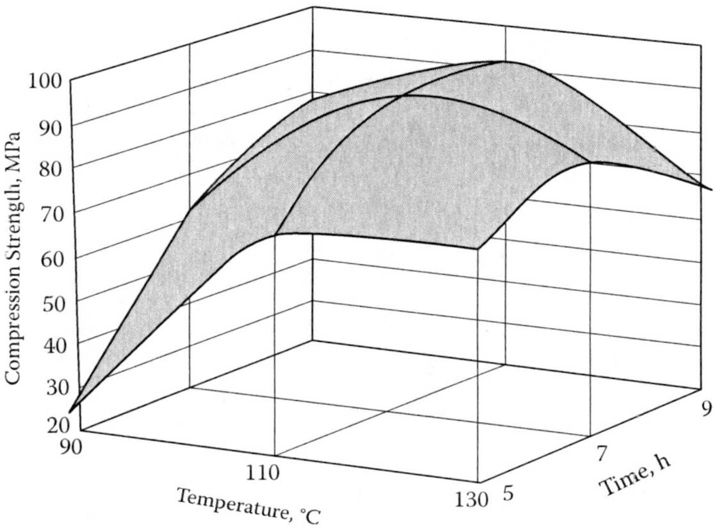

FIGURE 2.78 The compression strength σ_c dependence of vulcanization temperature mode. (Reprinted from O. Figovsky and D. Beilin, "Building Materials Based on Advanced Polymer Matrix: Review," *J. Scientific Israel Technological Advantages* 10, no. 3 (2008): 1–119. With permission.)

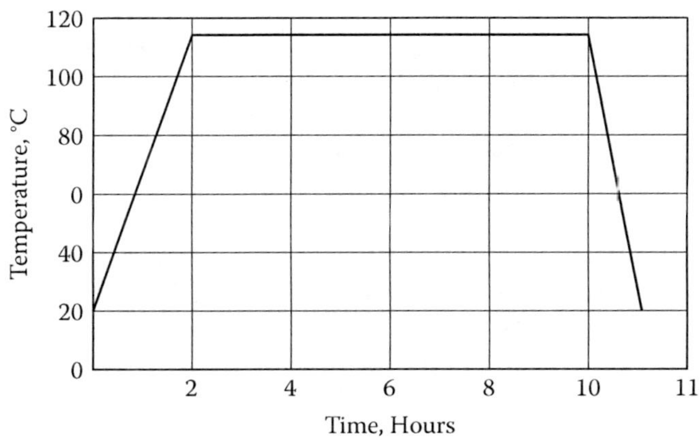

FIGURE 2.79 Recommended mode of heat treatment of RubCon's mix. (Reprinted from O. Figovsky and D. Beilin, "Building Materials Based on Advanced Polymer Matrix: Review," *J. Scientific Israel Technological Advantages* 10, no. 3 (2008): 1–119. With permission.)

At the present time, an accelerated sulfur vulcanization system is used for RubCon curing. This system consists of sulfurs as the structuring agent of vulcanization, tetramethylthiuram disulfide and 2-mercaptobenzothiazole as accelerators, and zinc oxide as the activator of this process.

This composition provides a vulcanizate with construction-material properties and high operational characteristics. However, these vulcanizing groups are not capable of providing an effective process for RubCon structurization at temperatures below 110°C.

Supplementary introduction of curing agents like diphenylguanidine (DPG) and synthetic fatty acids (SFA) allows a lower curing temperature of RubCon compositions due to formation of additional sulfide complexes, and decreases in the critical temperature required for accelerator action. With DPG and SFA, the acid–alkaline (pH) balance of the curing component interaction is not changed; any disturbance of the balance exerts an adverse effect on the kinetics of hardening.

Introduction of DPG alone will deoxidize the curing system due to its alkaline properties. The result is that structurization of the polymer will occur more quickly, violently, and at earlier stages, resulting in a sharp increase of interstructural defects of the composite, and hence deterioration of its physical–mechanical properties.

Introduction of SFA restores the acid–alkaline interaction of the curing group components and allows more effective hardening of the mixture. In addition, SFA reduces the viscosity of the composition, improves dispersion of the powder-like components, and promotes the formation of intermediate highly active compounds and their reaction with macromolecules of the polymer.

Thus, complex introduction of DFG and SFA in the RubCon mixture lowers the temperature of polymeric composition curing from 125°C to 80°C.

Another way of reducing the temperature of vulcanization of rubber binder is *regulation of the microstructure and viscosity of the dienic oligomer.* We know that liquid dienic rubbers are the macromolecules of oligomers with1,4-cis-, 1,4-trans-, and 1,2-vinyl- links with viscosities from 0.4 up to 10 Pa·c. It was found experimentally that production of "cold-cured" RubCon is possible if the content of 1,4-cis- links in a macromolecule of rubber is more than 65%, and the viscosity of the polymer is more than 17 Pa·c. Increasing the viscosity of rubber up to 7 Pa·c leads to lowering of the structurization temperature from 125°C to 60°C.

Simultaneously increasing in the content of 1,4-cis- links at reduction of the content of 1,2-vinil- links (up to 15%–20%) and increasing the viscosity allow vulcanization of the *RubCon* composition at room temperature (Figures 2.80 and 2.81).

METHOD OF THERMAL TREATMENT OF A PROTECTIVE COVERING BASED ON LIQUID POLYBUTADIENE BINDER BY ELECTRIC CURING [31]

An effective method for thermal treatment of a protective covering based on liquid polybutadiene binder is by electric curing using heating wire in ground coverings. A schematic diagram of the device for electric curing of the RubCon-type covering

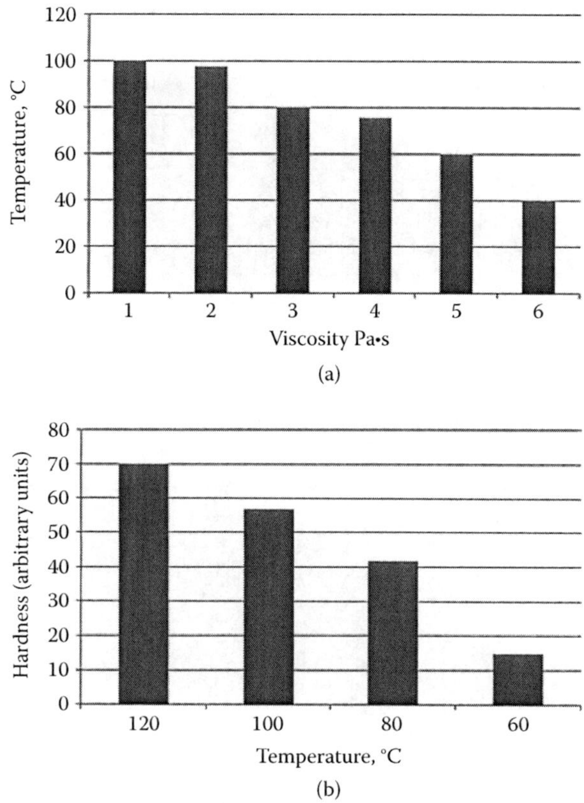

FIGURE 2.80 (a) Minimal temperature of vulcanization of RubCon composition based on diene oligomer, (b) relationship between temperature of vulcanization and hardness of the RubCon mixture. (Reprinted from O. Figovsky and D. Beilin, "Building Materials Based on Advanced Polymer Matrix: Review," *J. Scientific Israel Technological Advantages* 10, no. 3 (2008): 1–119. With permission.)

(usually called *liquid ebonite mixtures* (LEM) [32–34]) by heating wire is presented in Figure 2.82.

RubCon coverings are produced by vulcanization of the mixture, with the following component proportions in mass%:

Low-molecular oligobutadiene	8–11
Sulfur	3–6.5
Thiuram	0.3–0.7
Zinc oxide	1.5–5.0
Calcium oxide	0.3–0.6
Flue ash from thermal power plant	7–10
Filler	the balance

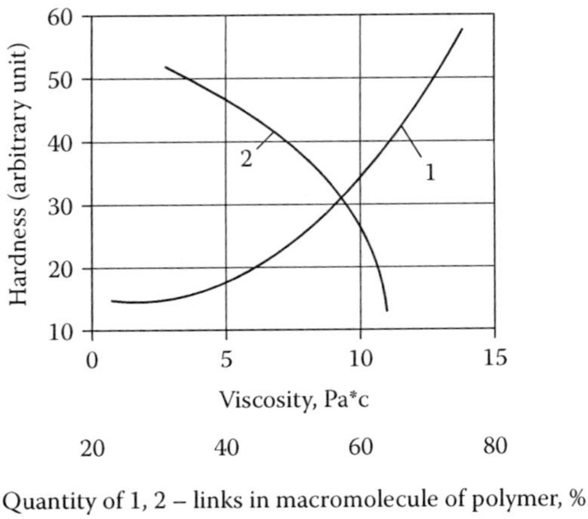

FIGURE 2.81 (1) The diene oligomer vulcanizate viscosity-hardness relationship after 5 hours curing at 100°C, (2) the quantity of 1,2-links in diene oligomer macromolecule influence on hardness of vulcanizate after 6 hours curing at 80°C. (Reprinted from O. Figovsky and D. Beilin, "Building Materials Based on Advanced Polymer Matrix: Review," *J. Scientific Israel Technological Advantages* 10, no. 3 (2008): 1–119. With permission.)

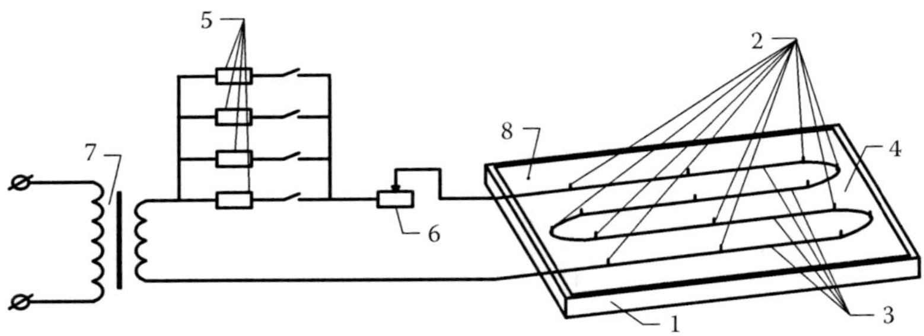

FIGURE 2.82 The diagram of the device for electric curing of RubCon's type covering with application of heating wire: (1) base layer, (2) dowels, (3) steel no insulated heating wires, (4) covering layer, (5) pull-up resistors, (6) rheostat, (7) step-down transformer. (Reprinted from Yu. Borisov, S. Matreninskiy, and R. Sapelkin, "Process of Heat Transfer during Vulcanization of the Concrete Based on Liquid Polybutadine Binder," *J. Scientific Israel Technological Advantages,* 11, no. 1 (2009): 15–22. With permission.)

Efficient temperature conditions for thermal treatment of the covering with heating wire have been set.

Thermal treatment consists of two stages (Figure 2.83):

- **First stage:** This stage lasts 55–65 minutes, providing that the temperature of the covering layer reaches 85°C–95°C for formation of the initial structural connections within the material. Here, the electrical linear load on the heating wire is 40 W/m.

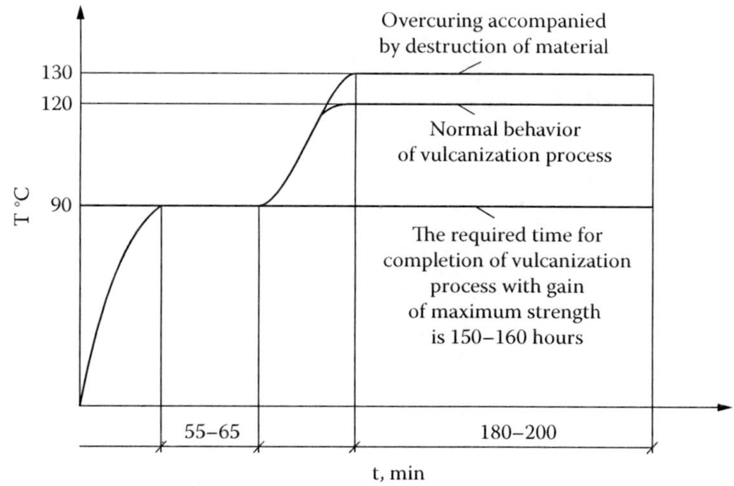

FIGURE 2.83 The diagram of thermal treatment of the covering within vulcanization by heating wire. (Reprinted from Yu. Borisov, S. Matreninskiy, and R. Sapelkin, "Process of Heat Transfer during Vulcanization of the Concrete Based on Liquid Polybutadine Binder," *J. Scientific Israel Technological Advantages,* 11, no. 1 (2009): 15–22. With permission.)

- **Second stage:** This stage lasts 180–200 minutes, during which the temperature of the covering layer is maintained at 115°C–125°C for completion of the vulcanization process and formation of a dense and homogeneous structure of the material. For maintenance of these conditions the electrical linear load on the heating wire is 60 W/m.

The rate of temperature increase must not exceed 1°C per minute to avoid destruction of the RubCon mixture during vulcanization due to intensive evaporation.

For analysis of temperature distribution over the volume of vulcanizate, there has been developed a mathematical model of the spreading of the temperature field within the vulcanizate depending on these boundary conditions: the type of heat-generating device, its temperature, and the ambient temperature.

Precise solution of the multidimensional problem of heat conductivity by analytical methods is very complicated and laborious. Therefore, an approximate finite difference method was developed based on the differential heat conductivity equation and boundary conditions. In this method, the temperature of the vulcanized section of the covering fragment was subdivided into elementary volumes of unit thickness because it is necessary to define the temperature field of the vulcanized.

The solutions of the system of linear equations represent the temperature field spreading within the vulcanizate. Graphical representation of the solution data is given in Figure 2.84.

Analysis of the temperature field allows a determination of the possibility of structure formation of the rubber concrete throughout the volume of vulcanized covering. The temperature field of the vulcanized covering has a small dispersion

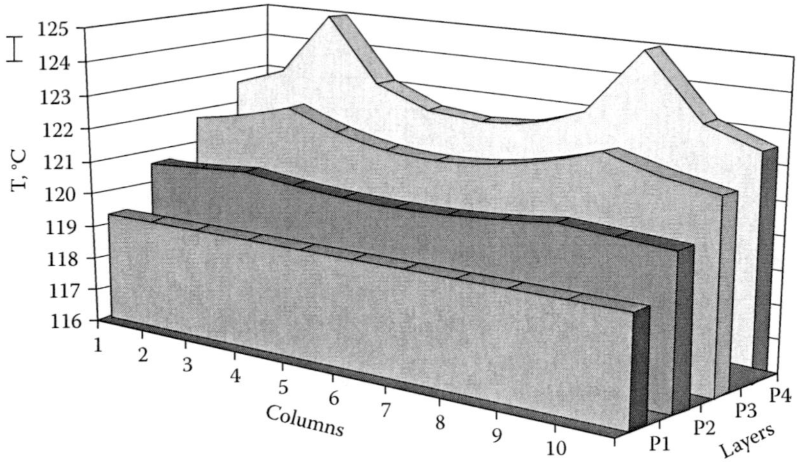

FIGURE 2.84 Temperature field in vulcanizate cross-section with the use of heating wire as heat-generating device. (Reprinted from Yu. Borisov, S. Matreninskiy, and R. Sapelkin, "Process of Heat Transfer during Vulcanization of the Concrete Based on Liquid Polybutadine Binder," *J. Scientific Israel Technological Advantages,* 11, no. 1 (2009): 15–22. With permission.)

in the given temperature range of 120°C–130°C (Figure 2.84). This provides uniform heating of the vulcanizate and normal behavior of structure formation processes.

The efficiency of vulcanization and structure formation of rubber concrete can be considerably increased by supplementary thermal treatment of the outside of the covering, thereby creating a more uniform thermal field along the surface and volume of the heated article.

As a result of analysis of heat-generating devices, the infrared radiation generator was selected as a supplementary source of thermal energy (Figure 2.85). This device produces directional heating of the surface and has a high coefficient of efficiency of 65%.

Experiments were conducted on a covering with a thickness of 8 mm. Heating wire was laid with the span of 20 mm under electrical load of 15 W/m. The mixture was heated up to 47°C (Figure 2.86, curve 3). Using supplementary infrared radiation on the outside of the covering, the field of specified temperature borders (118°C–125°C) was achieved throughout the volume of the material. Such temperature conditions were reached with a speed of 2°C/min without a failure of the material structure (Figure 2.86, curve 1). The temperature field appearing in the vulcanizate as a result of the combined thermal action of heating wire and infrared radiation provides uniform heating of the material and accelerated structure formation. The diagram of the temperature increase of the outside of the covering is shown in Figure 2.86, curve 2.

Figure 2.87 shows the temperature field distribution in the vulcanizate cross-section with heating wire as the heat-generating device, with supplementary infrared radiation on the outside surface.

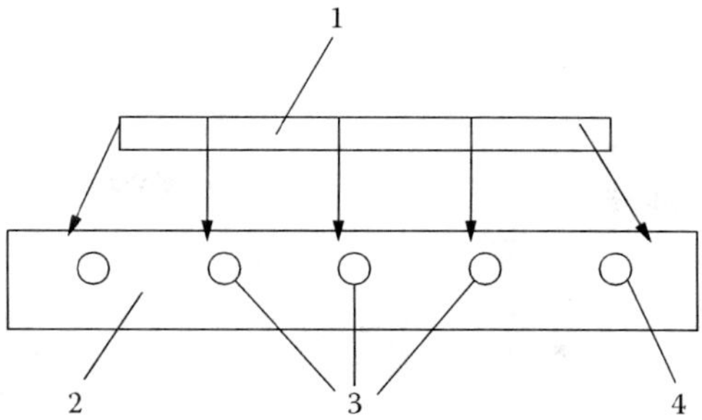

FIGURE 2.85 The layout of heating devices during complex heating of RubCon's type covering by heating wire and generator of infrared radiation: (1) generator of infrared radiation, (2) RubCon's type covering, (3) heating wire, (4) casing. (Reprinted from Yu. Borisov, S. Matreninskiy, and R. Sapelkin, "Process of Heat Transfer during Vulcanization of the Concrete Based on Liquid Polybutadine Binder," *J. Scientific Israel Technological Advantages,* 11, no. 1 (2009): 15–22. With permission.)

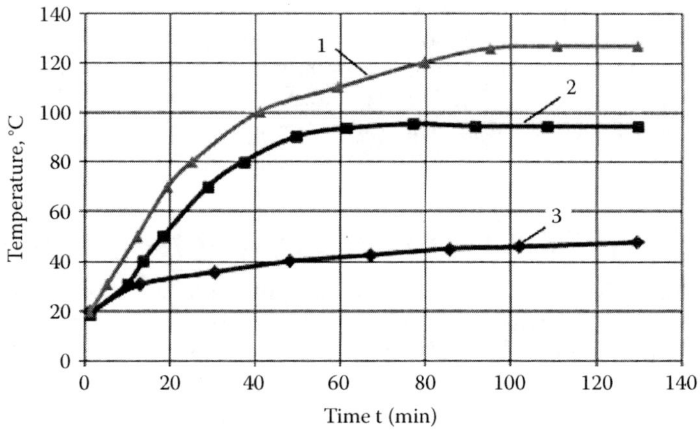

FIGURE 2.86 The diagrams of the rubber concrete temperature increase during different methods of thermal treatment: (1) heating wire combined with infrared radiation, (2) infrared radiation generator, (3) heating wire. (Reprinted from Yu. Borisov, S. Matreninskiy, and R. Sapelkin, "Process of Heat Transfer during Vulcanization of the Concrete Based on Liquid Polybutadine Binder," *J. Scientific Israel Technological Advantages,* 11, no. 1 (2009): 15–22. With permission.)

Summary

- Realization of a one-stage vulcanization process in which the temperature conditions are reached with a speed of 2°C/min without a failure of the material structure
- Increase of thermal field uniformity, which positively affects the strength characteristics of the rubber concrete
- Acceleration of the structure formation process

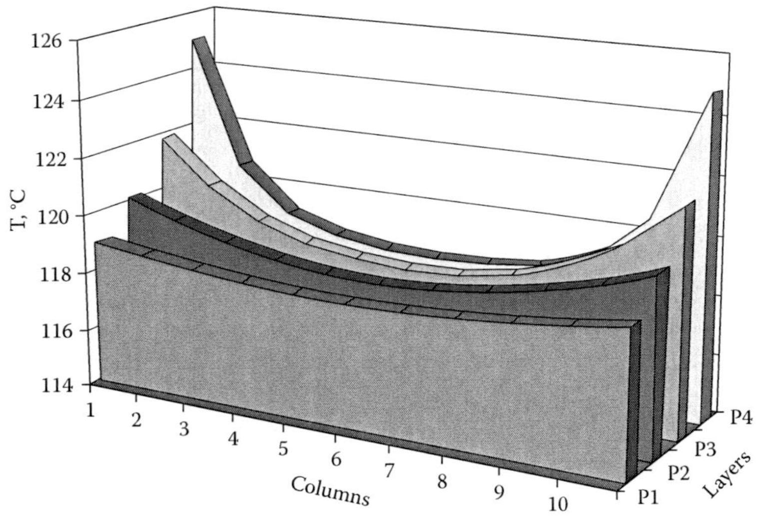

FIGURE 2.87 Graphical representation of temperature field distribution in vulcanizate cross-section with the use of heating wire as heat-generating device, with supplementary effect of infrared radiation on the outside surface. (Reprinted from Yu. Borisov, S. Matreninskiy, and R. Sapelkin, "Process of Heat Transfer during Vulcanization of the Concrete Based on Liquid Polybutadine Binder," *J. Scientific Israel Technological Advantages,* 11, no. 1 (2009): 15–22. With permission.)

PRODUCTION TECHNOLOGY OF FIBER-REINFORCED RUBCON

Introduction of fibers into composite materials influences not only their stress–strain characteristics but the manufacturing technology of these materials. The preparation process and workability of the mix depend on the percentage of fiber reinforcement (μ) and the relative length of the fibers.

The influence of agitation time (feeding of fibers into the concrete mix) on its strengthening properties was studied. Preparation of the fiber-reinforced RubCon mixture was performed with a high-speed laboratory mixer with speeds of 8.3 rpm and 16.6 rpm. Nonuniform distribution of reinforcing fibers and caking were observed at an agitation speed of 16.6 rpm. Therefore, an agitation speed of 8.3 rpm was used. The effect of agitation time or feeding of fibers into the mix on the bending strength of fiber-reinforced RubCon is illustrated in Figures 2.88 and 2.89. The bending strength of fiber-reinforced RubCon increases linearly with increases in the time during which fiber is fed into the mix. Bending strength then reaches a plateau and decreases (Figure 2.88). Notice that at fiber lengths of l/d = 40 and *l/d* = 70, the horizontal parts of the graphs with a constant bending strength value are greater than at lengths of *l/d* = 100 and *l/d* = 130.

It can be seen in Figure 2.88 that increasing the percentage of fiber reinforcement leads to an increase of agitation time. It is safe to assume that low values of bending strength of RubCon at short fiber-feeding times are related to formation of defects and pores as a result of nonuniform distribution of the fibers. Further increase in

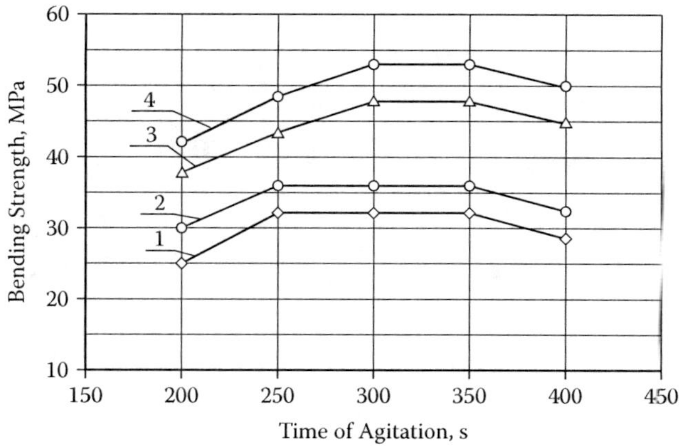

FIGURE 2.88 Bending strength of the fine-grained RubCon at μ = 2% depending on time of agitation (speed 8.3 rpm). (1) *l/d* = 40, (2) *l/d* = 70. (3) *l/d* = 100, (4) *l/d* = 130. (Reprinted from Yu. Borisov, S. Matreninskiy, and R. Sapelkin, "Process of Heat Transfer during Vulcanization of the Concrete Based on Liquid Polybutadine Binder," *J. Scientific Israel Technological Advantages,* 11, no. 1 (2009): 15–22. With permission.)

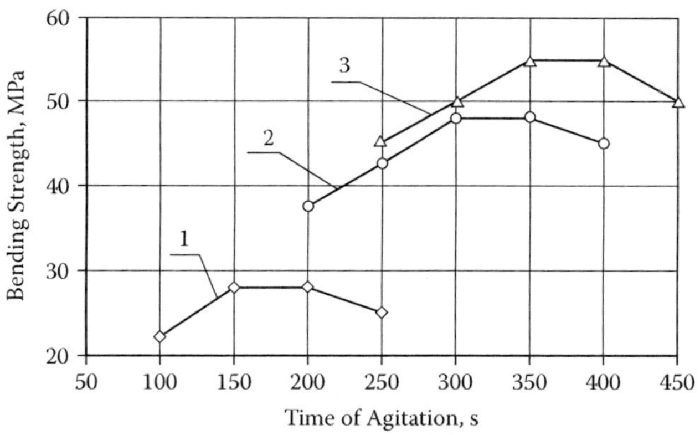

FIGURE 2.89 Dependence of bending strength of the fine-grained fiber reinforced RubCon on agitation time at speed 8.3 rpm, *l/d* = 100 and percent reinforcing: (1) μ = 2%, (2) μ = 2%, (3) μ = 3%. (Reprinted from Yu. Borisov, S. Matreninskiy, and R. Sapelkin, "Process of Heat Transfer during Vulcanization of the Concrete Based on Liquid Polybutadine Binder," *J. Scientific Israel Technological Advantages,* 11, no. 1 (2009): 15–22. With permission.)

fiber-feeding time leads to nonuniform distribution of fibers in a mixture, and formation of local supersaturation of fibers in condensed RubCon mixture parts and their absence in other parts, which results in a decrease in bending strength.

One of problems of dispersed reinforcement of composites is a decrease of mixture workability at an increased reinforcement percentage and ratio between length of fibers and their diameter. The workability criterion is the mobility of the mixture, which was estimated by slump of the standard cone.

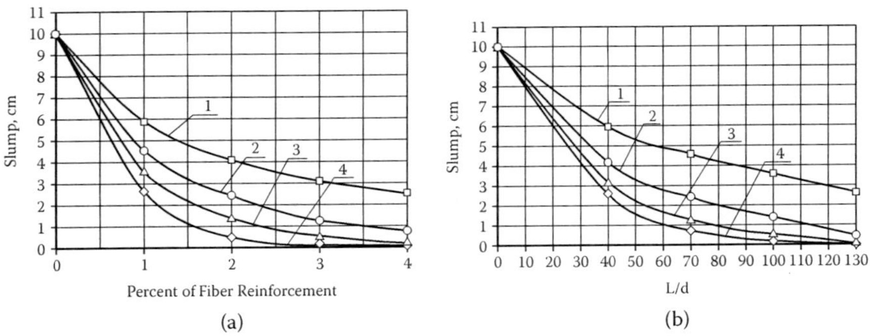

FIGURE 2.90 Mobility of RubCon mixture with *coarse* aggregate depending on: (a) percentage of fiber reinforcement at (1) $l/d = 40$, (2) $l/d = 70$, (3) $l/d = 100$, (4) $l/d = 130$; (b) ratio between length of fiber and it diameter at (1) $\mu = 1\%$, (2) $\mu = 2\%$, (3) $\mu = 3\%$, (4) $\mu = 4\%$. (Reprinted from Yu. Borisov, S. Matreninskiy, and R. Sapelkin, "Process of Heat Transfer during Vulcanization of the Concrete Based on Liquid Polybutadine Binder," *J. Scientific Israel Technological Advantages,* 11, no. 1 (2009): 15–22. With permission.)

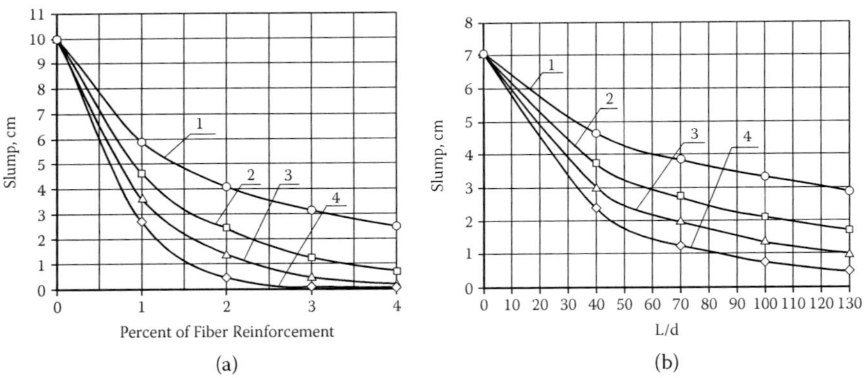

FIGURE 2.91 Mobility of RubCon mixture with *fine-grained* aggregate depending on: (a) percent of fiber reinforcement at (1) $l/d = 40$, (2) $l/d = 70$, (3) $l/d = 100$, (4) $l/d = 130$; (b) ratio between length of fiber and it diameter at (1) $\mu = 1\%$, (2) $\mu = 2\%$, (3) $\mu = 3\%$, (4) $\mu = 4\%$. (Reprinted from Yu. Borisov, S. Matreninskiy, and R. Sapelkin, "Process of Heat Transfer during Vulcanization of the Concrete Based on Liquid Polybutadine Binder," *J. Scientific Israel Technological Advantages,* 11, no. 1 (2009): 15–22. With permission.)

The mobility of a RubCon mixture with fine-grained and coarse aggregates depends on the percentage of fibers it contains and their relative length. This was obtained experimentally and is shown in Figures 2.90 and 2.91.

One can see from Figures 2.90 (a, b) and 2.91 (a, b) that the slump of standard cone decreases with an increase in reinforced fibers in the RubCon mixture with coarse and fine-grained aggregates. This is especially true at relative length of fibers above $l/d = 70$. At $\mu = 4\%$ and $l/d = 130$, the slump is practically equal to zero due to formation of a fiber skeleton of mixture components. Rigidity of this skeleton tends to increase in the process of growth of relative fiber length, which reduces the workability of the mixture.

Based on experiments, it can be concluded that application of fibers with a relative length of $l/d = 100$ is inefficient, especially at $\mu \geq 2\%$, because of the drastic reduction of the RubCon mixture workability and increased power inputs in the mixture preparation process.

Thus, the basic stages of fiber-reinforced RubCon manufacturing techniques are

- Preparation of formworks
- Preparation of a mix
- Placing of the mix into the formworks
- Compaction of the mix
- Vulcanization
- Extraction of finished articles from formworks

Preparation of the fiber-reinforced RubCon mix includes

- Washing and drying the filler and aggregates
- Production of fibers from metal cord waste
- Measuring of components

Fibers are obtained by chopping metal cords into appropriate length pieces and following deep cutting in a mixer. Agitation of components is done in the high-speed mixer at a speed of 16.6 rpm for 70 seconds. Rubber binder is prepared by mixing of liquid polybutadiene heated to 70±5°C with the components of curing groups and preheated fillers. Fine-grained and coarse aggregate are warmed to 70±5°C and are added to the mixture after preparation of the rubber binder at continuous agitation at a speed of 8.3 rpm for 200 seconds. Further fibers are fed into the RubCon mix at continuous agitation at a speed of 8.3 rpm for 300–350 seconds. The prepared mix is placed in warmed metal formworks and is compacted by a standard platform-type vibrator. The duration of vibration is about 120 seconds; evidence of sufficient compaction is the appearance of the binder on the mix surface and termination of air bubbling. It is necessary to apply a special layer to prevent finished articles from sticking to the formworks. It is possible to use silicon liquids, a solution of silicon-organic rubber in benzene. The rest of the manufacturing process of fiber-reinforced RubCon does not differ from production of plain RubCon (see previous sections in this chapter).

FIELD OF APPLICATION OF RUBCON [1,3,29,30]

The efficiency of RubCon structures and products in the construction industry is a result of its favorable combination of physical–mechanical and operational properties.

The high stress–strain characteristics and universal chemical resistance of concrete based on liquid rubber make possible to apply it in various chemically aggressive environments (subsoil waters, solutions of inorganic and mineral acids, alkalis, petroleum products, solvents, sewer drains, atmospheric precipitation, waste of industrial production, etc.) to make very rigid structures.

RubCon's outstanding properties have a multitude of applications that surpass the boundaries that limit conventional cementitious concrete. The following applications represent only a fraction of the numerous possibilities.

- **Industrial flooring:** RubCon's exceptional resistance to a wide range of corrosive chemicals—from acids to alkalis (unlike other products that resist only acids or bases but not both)—defines its usefulness in industrial floorings. Floors in factories, chemical laboratories, manufacturing plants, and so on will reap the benefits of RubCon by eliminating the need to resurface or replace [29].
- **Pickling, galvanic, electrolysis, plating baths:** RubCon effectively provides added wear resistance In equipment and structures operating in aggressive media. Pickling, galvanic, and electrolysis baths are some of the processes that can benefit from RubCon.
- **Containment structures:** Concrete vaults used to contain radioactive waste and burial containers for toxic and radioactive wastes can be constructed of RubCon to provide increased resistance to corrosive materials and chemicals.
- **Seismic reinforcement:** RubCon's high elasticity makes for a most impressive material in structures threatened by seismic activity. Where conventional concrete is brittle and tends to crack easily, RubCon's unique rubberlike binder offers a very high deflection rate, equipping the material with the freedom of movement necessary to rebound from seismic activity. As a result, highways, bridges, high-rise buildings, underground pipes, protective structures, and others constructed of RubCon perform favorably in earthquake conditions in comparison to conventional concrete structures.
- **Support foundations:** Stability, support, and flexibility are necessary standards for foundations that support tractable machinery. For example, pumps used to move massive amounts of liquids and solids require a foundation that will support their weight and movement. While conventional concrete provides support, it does not allow for flexibility in movement. RubCon's unique elastic properties satisfy both criteria.
- **Climate variations:** RubCon offers added endurance to structures battered by harsh weather conditions. Structures such as roadways, bridges, and so on, are adversely affected by temperature fluctuations and are in constant need of repairs. Once again, RubCon's favorable elastic property proves advantageous. Allowing for movement, RubCon can help to control structural cracking as a result of thermal stress brought about by constant shrinkage and expansion as materials react molecularly to hot and cold temperatures. Laboratory testing on ALNEC-PC samples revealed that after 1000 cycles of freezing and thawing, no cracks or weight loss could be reported. Even the finest cement concrete can withstand no more than 500 cycles.
- **Overlay applications:** In addition to its elasticity, RubCon is highly repellent to water. In fact, laboratory tests report that RubCon can be used as

an overlay in structures, dams, highways, and bridge decks to provide a durable wearing surface that is waterproof and, therefore, impervious to saltwater penetration.

- **Underground structures:** RubCon effectively offers protection against erosion caused by underground water. Minerals, acids, and alkalis found in underground water will aggressively attack individual components eventually.
- **Composite structures:** Application of two-layer RubCon–conventional concrete structures (with RubCon in the tension region) protects steel reinforcement from the effects of corrosive environments and creates a structure that operates under load without cracks. Research [30] proved that two-layer RubCon–concrete bending structure elements can replace prestressed concrete elements.
- **Ties (sleepers) for high-speed railways:** RubCon can greatly extend the life of concrete rail ties (sleepers) on high-speed railways and offers increased strength and vibration-absorbing and shock-resistant properties.
- **Civil engineering and infrastructure:**
  - Maritime structures (pilings, docks, and other exposed structures)
  - Bridge expansion joints and bridge deck overlays
  - Repairs made to load-bearing supports
  - Columns, beams, supports, skeleton structures in aggressive environments
  - Piers, viaducts for technological equipment into chemical industry buildings
  - Overflow areas, outside drainage pits
  - Sewer pipes for acid and alkali solutions
  - Vent tubes for wet acid gases

The following experimental adaptations of RubCon structures were done as part of our research:

- RubCon trough gutters were installed for sewer water drainage at a cattle-breeding farm in the Voronezh area. The previously installed reinforced concrete became unfit for use due to the influence of ammoniac steam corrosion. The new trough gutters measured 100 × 40 × 30. There were no visible changes in gutter surfaces after one year of operation. The control samples showed good chemical resistance (K_{xc} > 0,8) after 360 days of operation in a 25% solution of ammonia.
- The polymer cement flooring in the Voronezh radio components factory were replaced with RubCon floor tiles when the original flooring became unusable. Solutions of acids and alkalis are used in this shop. *RubCon* floor tiles measuring 250 × 250 × 20 mm were used. During three years of operation with regular exposure to acid and alkali solutions, no visible changes were detected.
- RubCon floor tiles measuring 500 × 500 × 60 mm, reinforced with a 100 × 100 mm mesh were laid in the Voronezh synthetic rubber plant, which uses acids and alkalis in the pumping process. The tiles were laid on a concrete substrate. There were no visible changes in the tiles after three years of

operation and constant exposure to concentrated solutions of sulfuric acid. Control tiles showed the highest chemical resistance (K_{xc} = 0.95–0.97) after 720 days of exposure in the same environment.

RubCon can be used in specialized structures instead of traditional building materials (reinforced concrete, steel, wood). Comparison of technical and economic parameters of RubCon with the most widespread kinds of corrosion-resistant concretes, in particular polymer concrete based on epoxy and polyester resins, showed that the cost of a useful loading unit of the RubCon structure is more than twice as effective as epoxy polymer concrete and 2.2 times as effective as polyester polymer concrete.

RubCon at operation in aggressive environments is 2.2–3.2 times as economically efficient as epoxy polymer concrete.

REFERENCES

1. Figovsky, O., and Beilin, D. "Building Materials Based on Advanced Polymer Matrix: Review," *J. Scientific Israel Technological Advantages* 10, no. 3 (2008): 1–119.
2. Figovsky, O., Potapov, Yu., Borisov, Yu., and Beilin, D. "RubCon: Technology of High Filled Composite Materials," RAPRA Technology Ltd., RubberChem 2002, *The Third International Rubber Chemicals, Compounding and Mixing Conference,* Munich, Germany, 2002, 21–31.
3. Potapov, Yu., Borisov, Yu., Barabash D., and Makarova T. *Effective Building Composites Based on Rubber Binders*, Voronezh, 2000 (in Russian).
4. Potapov, Yu., Figovsky, O., Borisov, Yu., Pinaev, S., and Beilin, D. "Stress–Strain State of Compressed Elements from Polymer Concrete," *J. Scientific Israel Technological Advantages* 4, no. 3-4 (2002): 20–24.
5. Potapov, Yu., Figovsky, O., Borisov, Yu., Chmyhov, V., and Beilin, D. "Influence of Temperature on Physical–Mechanical Characteristics of Polymer Concrete, *J. Scientific Israel Technological Advantages* 5, no. 1–2 (2003): 11–13.
6. Figovsky, O., Potapov, Yu., Makarova, T., and Beilin, D. "Load-Carrying Capacity of Polymer Concrete with Polybutadiene Matrix," *J. Scientific Israel Technological Advantages* 4, no. 1–2 (2002): 21–24.
7. Jurkov, S., *Physics of Strength and Plasticity*, Nauka, 1986 (in Russian).
8. Jartsev, V., "Physical-Technical Basis of Load-Carrying Capacity of Organic Materials in Parts and Structures," PhD thesis, Tambov, 1998 (in Russian).
9. Simonescu, C., and Oprea, C. *Mechanochimia compusilor macromoleculari*, Bucuresti, 1967.
10. Potapov, Yu., Figovsky, O., Borisov, Yu., Pinaev, S., and Beilin, D. "Joint Work of Reinforcement and Polymer Concrete Matrix," *J. Scientific Israel Technological Advantages* 4, no. 3–4 (2002): 14–19.
11. Borisov, Yu., Pinaev, S., and Savchenko, S. "Bond between Cauton and Non-Stretched Ribbed Reinforcement," *Ecological Bulletin of Black Earth Area* 11, Voronezh, 2001, 68–72 (in Russian).
12. Potapov, Yu., Figovsky, O., Borisov, Yu., Polikushkin, A., and Beilin, D. "Influence of Shear Force on the Behavior of Polymer Concrete Beams at Bend," *J. Scientific Israel Technological Advantages* 4, no. 3–4 (2002): 25–31.
13. Potapov, Yu., Borisov, Yu., Panfilov, D., Figovsky, O., and Beilin, D. "Research of Polymer Concrete Based on Low Molecular Polybutadiene, Part VI: Influence of Particle Size of Coarse Aggregates and Reinforcement Ratio on the Fibrous Polymer Concrete Strength," *J. Scientific Israel Technological Advantages* 6, no. 3–4 (2004): 67–70.

14. Potapo, Yu., Borisov, Yu., Panfilov, D., Figovsky, O., and Beilin, D. "Research of Polymer Concrete Based on Low Molecular Polybutadiene, Part VII: Strength of Continuously Reinforced Polymer Concrete with Various Kinds of Fibers," *J. Scientific Israel Technological Advantages* 6, no. 3–4 (2004): 71–74.
15. Parker, F. Jr. "Steel Fibrous Concrete for Airport Pavement Application." *International Conference on Concrete Pavement Design,* Purdue University, February 1977, 541–555.
16. Rabinovich, F. *Concrete with Dispersed Reinforcement.* Moscow: Stroyizdat, 1989 (in Russian).
17. Figovsky, O., Beilin, D., Blank, N., and Potapov, Yu. "Development of Polymer Concrete with Polybutadiene Matrix." *J. Cement and Concrete Composites* 18, no. 6 (1996): 437–444.
18. Potapov, Yu., Figovsky, O., Borisov, Yu., Pinaev, S., and Beilin, D. "Creep of Polymer Concrete at Compressive Loading," *J. Scientific Israel Technological Advantages* 5, no. 1–2 (2003): 1–10.
19. Potapov, Yu., Mailyan, L., Borisov, Yu., and Pinaev, S. "Creep of the Concretes Based on Rubber Matrix Substantion at Compression," *Russian Academy of Architecture and Construction, Bulletin of Construction Sciences* 3 (2000): Moscow, 215–226 (in Russian).
20. Ivanov, A., Potapov, Yu., and Alimov, S. "About the Equation of Non-linear Creep of Some Plastics and Wood," *Proceedings of Higher Institutes of Learning* no. 6 (1968), (in Russian).
21. Chmyhov, V. "Resistance of Rubber Concrete to Action of Aggressive Environments," doctoral thesis, Voronezh, 2002 (in Russian).
22. Potapov, Yu., Borisov, Yu., Chmyhv, V., and Beilin D. "Research of Polymer Concrete Based on Low Molecular Polybutadiene, Part VIII: Chemical Resistance of Polymer Concrete," *J. Scientific Israel Technological Advantages* 7, no. 1–2 (2005): 72–78.
23. Figovsky, O., Chmyhov, V., and Beilin, D. "Corrosion Resistance of Rubber Concrete," *Open Corrosion Journal* no. 3 (2010): 28–37, http://www.benthamscience.com/open/tocorrj/articles/V003/28TOCORRJ.pdf.
24. Potapov, Yu., Borisov, Yu., and Chmyhov, V. "Increasing of Durability of Rubber Concrete." In *The New Scientific Ways of Construction Material Engineering*, Part II, 2005, Belgorod, Russia, 88–94 (in Russian).
25. Perekalsky, O. "Constructional Composites Based Polybutadiene Oligomers for Radiation Protection," doctoral thesis, Voronezh, 2006 (in Russian).
26. Surovtsev, I., Borisov, Yu., Perekalsky, D., Figovsky, O., and Beilin, D. "Attenuation of γ-Radiation by Polymer Concrete and Its Resistance to Radioactive Radiation," *ICPIC 2007,* 12th International Congress on Polymers in Concrete, Chuncheon, Korea.
27. Solomatov, V., and Seljaev, V. *Chemical Strength of Composite Building Materials.* Moscow: Stroyizdat, 1987 (in Russian).
28. V'yugov, P., Goncharov, K., Dementii, V., and Mandrichenko, A. "Attenuation of Gamma-Radiation by Concrete and Some Naturally Occurring Materials," *J. Atomic Energy (Historical Archive)* 10, no. 1 (1961): 78–80.
29. Figovsky, O., Borisov, Yu., and Beilin, D. "Rubber Concrete for Industrial Floors." In *Proceedings 6th International Colloquium, "Industrial Floors 07,"* Germany, vol. 1, 2007, 155–163.
30. Borisov, Yu., Polikutin, A., and Nguyen Phan Duy. "Research of Reinforced Two-Layer Beams Made from Conventional and Rubber Concretes," *J. Scientific Israel Technological Advantages* 14, no. 2 (2012): 5–23.
31. Borisov, Yu., Matreninskiy, S., and Sapelkin, R. "Process of Heat Transfer during Vulcanization of the Concrete Based on Liquid Polybutadine Binder," *J. Scientific Israel Technological Advantages,* 11, no. 1 (2009): 15–22.

32. Pushkarev, Yu., and Figovsky, O. "Protective Vulcanized Coatings on the Base of Oligobutadienes without Functional Groups," *Corrosion Reviews* 7, no. 1 (1999): 33–46.
33. Pushkarev, Yu., and Figovsky, O. "Protective Ebonite Coatings on the Base of Oligobutadiene," *Anti-Corrosion: Methods and Materials* 46, no. 4 (1999): 261–267.
34. Figovsky, O., and Blank, N. "Liquid Ebonite Mixtures for Anticorrosive Coverings," *International Conference Corrosion in Natural and Industrial Environments: Problems and Solutions,* Grado, Italy, May 23–25, 1995, 593–599.

3 Polymer Concrete Based on an Organo–Silicate Matrix

Silicate polymer concrete (SPC) is used in the repair of foundations, for the manufacture of storage tanks for hot and cold acids, equipment lining, reservoirs, and building structures, and as the structural material in vessel heads, vaults, and diaphragms. SPC is of particular interest in precast structures in production of pickling baths for metallurgical plants.

SPC is applied for monolithic chemical resistance and highly filled floorings, production of load-bearing structures in acid media (columns, beams, foundations, etc.), and in other fields where sound building materials are required with resistance to acid, water, and heat.

It is well known that SPC consists of a binder, a hardener, a polymeric additive, a filler, and an aggregate. Water-soluble sodium silicate or potassium glass with density (1.38–1.4)10 g/mm^3 applies as a binder. Technical sodium silicofluoride is used in most cases as a hardener. Fillers and aggregates are natural or artificial materials with acid resistance of no less 90%, in particular: diabases, basalts, granites, andesites, and so on. Silicate polymers can be prepared in the form of mastics, solutions, or concrete depending on grain size distribution. SPC has many important consumer properties: high density, fire resistance, durability, and acid resistance. However, these concretes have grave practical disadvantages: small application life and large shrinkage deformation.

A significant increase in silicate matrix strength and density may be achieved by incorporation of special liquid organic alkali-soluble silicate additives such as tetrafurfuryloxisilane (TFS), which in the hydration process forms the active nanosilica particles SiO_2, orthosilicic acid, furfuryl alcohol (FA), and other chemical species making oligomer nanofilms on a surface of silicate matrix grains. This effect is attributable to the hardening of contacts between silicate binder gel globes and modification of the alkaline component due to "inoculation" of the furan radical. Introduction in composition of these additives makes concrete practically impermeable to various acid solutions.

The optimal concrete composition with increased strength, chemical resistance in aggressive environments, good workability, and high density was obtained.

The use of prefabricated SPS structures and articles has several advantages over monolithic structures, including high quality of manufacturing and repairability allowing replacement of individual structural elements. In this regard, the development of acid- and temperature-resistant adhesive compounds and study of the adhesive strength of the butt joints of prefabricated SPS structures are actual problems.

The important quality of SPS products and structures is the resistance to occurrence and development of cracks. Results of experimental and theoretical study of fracture and crack resistance of SPC are discussed below.

OPTIMAL COMPOSITION OF SILICATE POLYMER CONCRETE

Influence of Liquid Glass and Monomeric Additive Content on SPC Fluidity, Harshness, Workability, and Strength

It is apparent that the strength of SPC as a structural material depends on the ratio of binder to aggregate at conservation of technological plasticity of a mix. The influence of liquid glass and the monomeric additive content on fluidity and harshness of SPC mixes was investigated. Results of the experiments are presented in Table 3.1.

One can see that a small change in the liquid glass quantity drastically changes the technical characteristics of SPC mixes. Reduction of the quantity of binder by 15% reduces fluidity of the mix by 2.5 times and increases its harshness by almost 5 times.

Influence of the monomeric additive (TFS) on the technological properties of SPC was investigated with plastic mix (content of liquid glass is 11%). Experimental results are illustrated in Table 3.2.

Results of the research show that introduction of TFS increases the harshness of SPC mixes. Regarding the workability of the investigated compositions, the SPC mix can be classified as "plastic" with content of liquid glass of 11% and monomeric additive of 3%." The influence of the quantity of liquid glass on SPC strength was defined both for compositions with the monomeric additive and without it.

The investigated SPC compositions are shown in Table 3.3. Results of experiments are shown in Figure 3.1 and Table 3.4.

One can recognize that the strength and density of SPC compositions increase with reductions in the liquid glass content in the entire range of its change. So, a

TABLE 3.1
Influence of Liquid Glass Content on Fluidity and Harshness of SPC Mixtures

Liquid Glass Content (%)	Fluidity of SPC (mm);	Harshness of the SPC Mixture (sec)	Kind of SPC Mix
13	150	5	High workability concrete
12	120	10	
11	60	23	Plastic concrete
10	0	30	Dry mix concrete

Source: Reprinted from O. Figovsky, D. Beilin, and Yu. Zemlyanushnov, "Fracture and Crack Resistance of Silicate Polymer Concrete," *Journal Scientific Israel Technology Advanced* 14, no. 4 (2012): 38–48. With Permission.

TABLE 3.2
Influence of the TFS Additive on Fluidity and Harshness of SPC Mixtures

TFS Additive Content (%) of Liquid Glass	Fluidity of the SPC Mixture (mm)	Harshness of the SPC mix (sec)
Without additive	60	23
2	40	24
3	20	28
6	0	32

Source: Reprinted from O. Figovsky, D. Beilin, and Yu. Zemlyanushnov, "Fracture and Crack Resistance of Silicate Polymer Concrete," *Journal Scientific Israel Technology Advanced* 14, no. 4 (2012): 38–48. With Permission.

TABLE 3.3
Composition of SPC Mixtures

Composition Number Components	1	2	3	4	5	6	7	8
	Content (% of mass)							
Liquid glass	10.00	11.00	12.00	13.00	10.00	11.00	12.00	13.00
Na_2SiO_6.	1.50	1.70	1.80	2.00	1.50	1.70	1.80	2.00
Diabase flour	18.00	18.00	18.00	18.00	18.00	18.00	18.00	18.00
Quartz sand	28.00	28.00	28.00	28.00	28.00	28.00	28.00	28.00
Granite chippings	42.50	41.30	40.20	39.00	42.20	40.97	39.84	38.61
TFS	—	—	—	—	0.30	0.33	0.36	0.39

Source: Reprinted from O. Figovsky, D. Beilin, and Yu. Zemlyanushnov, "Fracture and Crack Resistance of Silicate Polymer Concrete," *Journal Scientific Israel Technology Advanced* 14, no. 4 (2012): 38–48. With Permission.

decrease of the liquid glass content by 3% (mixes 1, 4, and 5, 8) leads to an increase in SPC strength of approximately 25%. It should be emphasized that introduction of 0.3% TFS in SPC compositions increases the strength and density of the material by approximately 50% in the entire range of liquid glass consumption. It is believed that the increase in strength and density of the SPC mix with a reduced content of liquid glass is connected with the thickness of the binder film that envelops the filler's large grains. As this takes place, the thinner the film, the better its adhesive bonding ability and, hence, the better the strength of the concrete as a whole.

Optimization of SPC Composition [1, 2]

Application of plastic SPC mixtures allows the production of structures with any configuration and conditions for concrete placement. This optimization of SPC

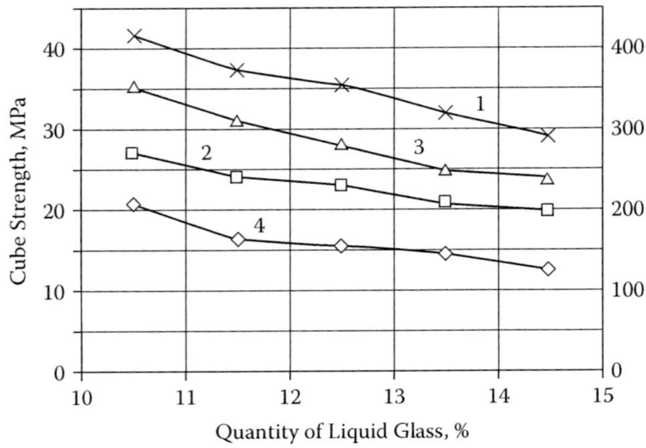

FIGURE 3.1 Change of strength and density of SPC depending on the consumption of the liquid glass: (1) change of SPC strength with the TFS additive, (2) same without the additive, (3) change of US velocity into SPC mix with the TFS additive, (4) same without the additive. (Reprinted from O. Figovsky and D. Beilin, "Optimal Composition, Strength, and Chemical Resistance of Silicate Polymer Concrete," *Proceedings of ICPIC 2010: 13th International Congress Polymers in Concrete*, February 10–12, 2010, Madeira, Portugal. With permission.)

TABLE 3.4
Influence of the Liquid Glass Content on Strength of SPC Compositions

Composition Number	1	2	3	4	5	6	7	8
Ultimate compression strength MPa	27.0	25.8	23.7	21.7	40.1	38.0	35.4	32.4

Source: Reprinted from O. Figovsky, D. Beilin, and Yu. Zemlyanushnov, "Fracture and Crack Resistance of Silicate Polymer Concrete," *Journal Scientific Israel Technology Advanced* 14, no. 4 (2012): 38–48. With Permission.

composition is dependent upon a quantity of liquid glass of 11%–11.5% and 3% of the monomeric additives FA or TFS (based on the weight of the liquid glass).

For optimization of SPC composition, the influence of three factors was studied.

- x_1 = Weight ratio of liquid glass to filler
- x_2 = Content of sand in a mix of fillers
- x_3 = Content of liquid glass in a mix

The effectiveness functions were the compressive strength of the SPC mix at 28 days, $\hat{Y}_{st}$, and the harshness of the SPC mix, $\hat{Y}_{hr}$, with the provision that harshness is not more than 30 seconds. The following regression equations were obtained

$$\hat{Y}_{st} = 37.6 - 5.7x_1 - 1.5x_2 + 17x_1x_3 \quad (3.1)$$

$$\hat{Y}_{hr} = 23 - 4.5x_1 - 15x_3 + 7.4\, x_1^2 + 4.4x_2^2 \tag{3.2}$$

by means of which the optimized composition of SPC was found (Table 3.5).

The physical–mechanical properties of the optimum SPC compositions are illustrated in Table 3.6. It should be noted that the SPC mix modified with TFS has higher strength indexes and greater deformability. Study of shrinkable deformations

TABLE 3.5
Optimal Composition of SPC

Composition Number	1	2	3
Components		**Content (% of mass)**	
Liquid glass ($1.4*10^{-3}$ g/mm^3)	11.23	11.23	11.23
$3Na_2SiF_6$.($2.7*10^5$ mm^2/g)	1.68	1.68	1.68
Diabase flour ($2.4*10^5$mm^2/g)	20.06	20.06	20.06
Quartz sand	26.71	2 26.71	26.71
Granite chippings (5–10 mm)	40.32	30.98	30.98
FA	—	0.34	—
TFS	—	—	0.34
Total	100.00	100.00	100.00

Source: Reprinted from O. Figovsky, D. Beilin, and Yu. Zemlyanushnov, "Fracture and Crack Resistance of Silicate Polymer Concrete," *Journal Scientific Israel Technology Advanced* 14, no. 4 (2012): 38–48. With Permission.

TABLE 3.6
Physical–Mechanical Properties of Optimal SPC Compositions

		Optimal Composition		
Index	**Unit**	**1**	**2**	**3**
Cube strength	MPa	20–25	28	36–41
Prism strength	MPa	—	20–22	30–35
Tensile strength	MPa	1.5	3.2	4.1
Tensile strength at bend	MPa	—	6	10
Elasticity modulus	MPa	$(1.6–.1.7)10^4$	$(1.9–2.1)10^4$	$(2.4–2.6)10^4$
Poisson coefficient		—	0.23	0.21
Longitudinal strain	mm/m	—	$(140–150)10^{-5}$	$(150–165)10^{-5}$
Transverse strain	mm/m	—	$(30–40)10^{-5}$	$(55–60)10^{-5}$
Toughness	kJ/m^2	—	2.3	5.4
Coefficient of thermal expansion	1/°C	—	$8*10^{-6}$	$8.7*10^{-6}$
Shrinkage at hardening (after 28 days)	%	0.39	0.22	0.06

Source: Reprinted from O. Figovsky, D. Beilin, and Yu. Zemlyanushnov, "Fracture and Crack Resistance of Silicate Polymer Concrete," *Journal Scientific Israel Technology Advanced* 14, no. 4 (2012): 38–48. With Permission.

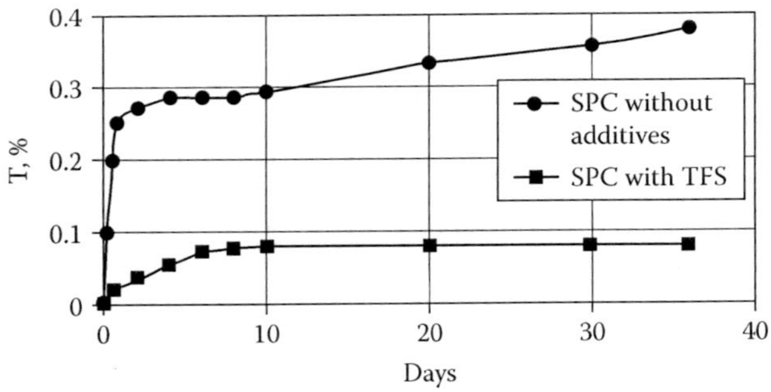

FIGURE 3.2 Change of shrinkable deformations T (%) during hardening of SPC. (Reprinted from O. Figovsky and D. Beilin, "Optimal Composition, Strength, and Chemical Resistance of Silicate Polymer Concrete," *Proceedings of ICPIC 2010: 13th International Congress Polymers in Concrete*, February 10–12, 2010, Madeira, Portugal. With permission.)

has shown that introduction of monomeric additives leads to a significant reduction in shrinkage. Use of 3% TFS (3% of the weight of liquid glass) reduces shrinkable deformations up to 0.06%. The greater shrinkage of the SPC mix without the additive take place in the first day (up to 70%), then it proceeds slowly and stops for 32 to 35 days (Figure 3.2).

SPC mix with monomeric additives is characterized by an exponential increase in shrinkage deformations. Formation of the SPC structure is accompanied by intensive compression of liquid glass gel by capillary forces of intermicellar liquid. The compression of the gel causes maximum shrinkage deformation from the very beginning of hardening. Introduction of FA or TFS considerably reduces the influence of capillary forces due to a decrease of liquid tension into capillaries [3,4].

Chemical Resistance and Durability of Silicate Polymer Concrete

It is known that an elevated permeability of usual SPC based on liquid glass is intimately connected with the big porosity and filtering defects. Because of this, application of commonly used SPC requires a leak proof and chemically stable lining of expensive materials (lead, polyisobutylene, etc.). Filtration and diffusive permeability of SPC with monomeric additives (TFS, FA) in aggressive environments were studied.

Filtration Permeability

To study filtration permeability, the compositions of SPC samples were accepted in accordance with the optimal composition given in Table 3.5. Water was used as a filtration liquid because it did not influence consolidation of the concrete. This allowed determinations about the presence or absence of filtering defects in investigated compositions and estimation of the influence of additives on filtration permeability. For SPC compositions tested with water, a filtration of acids is absent, as the effect of consolidation of the material in this case is shown.

TABLE 3.7
Filtration Permeability of the SPC Compositions

Composition	Permeability at Pressure, MPa
1	0.1
2	0.6
3	0.6

Source: Reprinted From O. Figovsky, D. Beilin, And Yu. Zemlyanushnov, "Fracture And Crack Resistance of Silicate Polymer Concrete," *Journal Scientific Israel Technology Advanced* 14, no. 4 (2012): 38–48. With permission.

Hydrostatic pressure was accepted as being 0.3–0.4 MPa. This value corresponded to application of SPC as a structural material in pressure tanks, bottom evaporative surface condensers, storage of acids, pickling baths, and so on. Maximum overflow pressure equaled 0.6 MPa. The greatest pressure at which water did not leak out through SPC samples and "sweating" of samples was not observed was accepted as a degree of waterproofing of the composition. Results of the test are presented in Table 3.7.

Data show that all SPC samples with monomeric additives withstand the pressure of water at 0.6 MPa, whereas samples without the additive (composition 1) began leaking after 1–1.5 hours of testing at an overflow pressure of 0.1 MPa. Thus, it is obvious that the additives TFS and FA eliminate filtering defects in the form of microcracks and conducting channels on a border "gel filler."

Diffusive Penetration

The durability of SPC structures in aggressive environments depends on the diffusion rate of chemically active reagents into a material. Fluid penetration is realized through imperfections in molecular packing of the binder and it is accelerated at an increase of temperature and pressure. In this connection, definition of a diffusion coefficient (as a key parameter of a liquid carryover) becomes the important problem.

It is known that a large number of aggressive liquids are used in the form of water solutions, and thus study of water diffusion in SPC is important. We investigated diffusive penetration of SPC in the neutral water environment, which is the most aggressive reagent for liquid glass compositions. With knowledge of a diffusion coefficient value for a time frame it is possible to estimate the influence of the monomeric additive and to establish a maximum allowable concentration of an aggressive environment. Results of tests of compositions 2 and 3 are illustrated in Figure 3.3.

It can be seen that the process of diffusive penetration of SPC compositions modified with FA after 25 days remains practically constant, unlike the composition 2, modified by TFS. The diffusive coefficients obtained by a sorption method were calculated for two periods of activity of the water media: 7 and 30 days. Results of the calculation are given in Table 3.8.

Therefore, introduction of modifiers from furan row in SPC composition allows reduction of the speed of liquid transfer velocity.

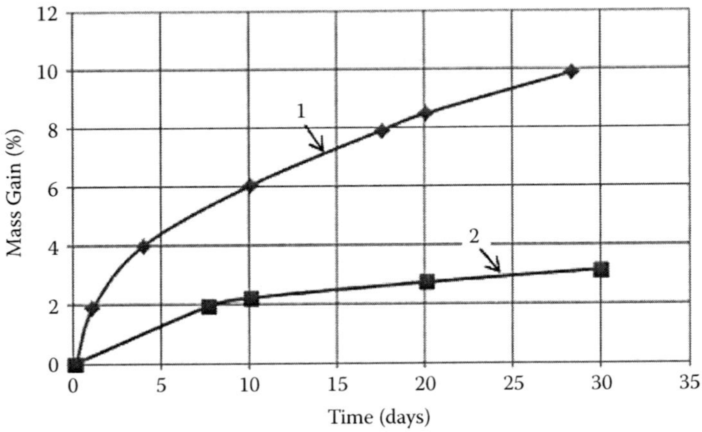

FIGURE 3.3 Change of weight of SPC samples in water: (1) composition modified by FA, (2) same but by TFS. (Reprinted from O. Figovsky and D. Beilin, "Optimal Composition, Strength, and Chemical Resistance of Silicate Polymer Concrete," *Proceedings of ICPIC 2010: 13th International Congress Polymers in Concrete*, February 10–12, 2010, Madeira, Portugal. With permission.)

TABLE 3.8
Diffusive Coefficients Water in SPC Compositions

Composition	Diffusion Coefficient $10^{-5}mm^2/sec$	
	After 30 days	After 7 days
8.91	8.77	2
0.25	0.74	3

Source: Reprinted from O. Figovsky, D. Beilin, and Yu. Zemlyanushnov, "Fracture and Crack Resistance of Silicate Polymer Concrete," *Journal Scientific Israel Technology Advanced* 14, no. 4 (2012): 38–48. With Permission.

Influence of Monomeric Additives on Chemical Resistance of SPC Compositions

Corrosion stability of SPC in high-concentration acid media permits its use widely in anticorrosive techniques. However, the corrosion resistance of SPC in alkalis and in weak water solutions of acids is not enough. The effect of corrosion in such environments is shown in deposits of salts in pores, cracks, and other voids. Crystallization of these salt deposits, introduced by the corrosion environment or formed by chemical reaction of the material and environment, leads to destruction of a silicate composition.

Highly diffusive penetration of SPC can be drastically lowered by active additives that are capable of raising SPC density, and, therefore, its corrosion resistance. As indicated previously, the most effective results were obtained by introduction of FA or TFS additives. Addition of monomeric additives to silicate composition improves the physical–mechanical characteristics and chemical resistance of silicate compositions due to improvement in the quality of silicate bonds and better adhesion between the binder and coarse filler [1]. In other words, the influence of the monomeric additives is conditioned by consolidation of liquid glass gel during hardening and modifications of alkaline components due to "inoculation" of furan radicals.

The consolidation mechanism of SPC compositions containing these additives can be considered as resulting from two processes: decompression of silicon acid gel due to a decrease in the surface tension of intermicellar liquid and polymerization under the influence of acid solutions.

Considering FA and TFS as consolidation additives, it should be noted that penetration of the SPC compositions modified by FA strongly depends on concentration of an aggressive environment. The polymerization rate of FA in the silicate media under the action of aggressive environments depends on pH, especially in the first reaction stages. Corrosion of silica by hydrofluoric acid occurs rather quickly and leads to loosening of the SPC composition and an increase of diffusive penetration. Consequently, application of the consolidation additive FA is desirable at pH ≤ 5.

The mechanism of TFS action as a consolidation additive in these environments is a little different: there is a hydrolysis of TFS with formation of an orthosilicic acid and FA [4]. Inorganic cement fills the saddles between globules of silica gel. An increase of the contact area leads to significant growth of durability, and in so doing, compensates for strength reduction due to etching a matrix by hydrofluoric acid. In that way, TFS essentially reduces the diffusive penetration of water and weak acids and allows use of SPC for reinforced structures without special protection of a reinforcement.

Chemical Resistance of SPC in a Faintly Acid Environment

SPC penetration in weakly acid environments has been studied for optimal composition including consolidation monomeric additives FA or TFS (Table 3.5). Concentration of aggressive environments included sulfuric and hydrochloric acid solutions applied for pickling of metals. Corrosion stability of SPC was estimated by change of compression strength of the SPC samples during 3–18 months exposure in the environment with 3-month intervals. Corrosion resistance coefficients of SPC compositions including FA and TFS additives are given in Table 3.9.

The experimental data allowed us to obtain the analytical interrelationship of chemical stability coefficients of the SPC compositions modified by TFS and FA additives in hydrochloric acid, $K_{(HCl., FA)}$, $K_{(HCl, TFS)}$, and sulfuric acid, $K_{(H2SO4, FA)}$, $K_{(H2SO4, TFS)}$, environments and their concentration and time of exposure:

$$K_{(HCl., FA)} = 0.99 + 2.19*10^{-3}\,C - 4.36*10^{-3}\tau + 5.07*10^{-4}C\tau \tag{3.3}$$

$$K_{(H2SO4, FA)} = 0.95 + 8.03*10^{-3}\,C - 0.13*10^{-3}\tau + 1.1*10^{-4}C\tau \tag{3.3a}$$

TABLE 3.9
Corrosion Resistance Coefficients of SPC Compositions

		SPC with TFS		SPC with FA	
Concentration of Acid (%)	**Time of Exposure (months)**	H_2SO_4	HCl	H_2SO_4	HCl
1		0.97	0.92	0.96	0.89
2		0.99	0.97	0.89	0.98
5	3	1.00	0.98	0.93	0.97
10		1.02	1.03	0.97	0.98
20		1.04	1.06	1.02	1.03
1		0.96	0.94	0.88	0.90
2		0.97	1.01	0.91	0.94
5	6	1.01	1.03	0.92	0.96
10		1.05	1.05	1.01	1.03
20		1.08	1.11	1.10	1.05
1		0.93	0.88	0.91	0.97
2		0.96	0.96	0.90	0.91
5	12	1.03	1.02	0.90	1.01
10		1.06	1.07	0.98	1.04
20		1.12	1.17	1.08	1.08
1		0.89	0.86	0.83	0.85
2		0.92	1.03	0.87	0.88
5	18	1.02	1.04	0.91	1.03
10		1.05	1.06	1.02	1.05
20		1.12	1.10	1.07	1.06

$$K_{(\text{HCl., TFS})} = 0.91 + 7.37*10^{-3}\,\text{C} - 3.62*10^{-3}\tau + 2.86*10^{-4}C\tau \tag{3.4}$$

$$K_{(\text{H2SO4, TFS})} = 0.93 + 4.21*10^{-3}\,\text{C} - 2.6*10^{-3}\tau + 6.1*10^{-4}C\tau \tag{3.4a}$$

where C is the solution strength (%) and τ is the time of exposition (months).

By taking $K \geq 0.8$, we can forecast the SPC composition service life in aggressive acid environments of various concentration.

Results concerning the influence of the kind of monomeric additive on SPC corrosion stability is shown in Table 3.10.

Adhesion Strength of Joints of Precast Silicate Polymer Concrete Structural Members [5]

New Adhesive Composition for Joint Grouting of Load-Bearing Prefabricated SPS Structures

The new adhesive compositions were used for joint grouting of reinforced SPS precast structures. The composition formulation is given in Table 3.11.

TABLE 3.10
Ultimate Compressive Strength of SPC versus Additive Type and the Corrosion Environment

	Aggressive Media		
Type of Additive	**Absence**	**2% H_2SO_4**	**2% HCl**
TFS	33.8	36.3	34.6
FS	25.3	25.0	24.8

Source: Reprinted from O. Figovsky, D. Beilin, and Yu. Zemlyanushnov, "Fracture and Crack Resistance of Silicate Polymer Concrete," *Journal Scientific Israel Technology Advanced* 14, no. 4 (2012): 38–48. With Permission.

TABLE 3.11
Adhesive Composition Formulation

	Composition				
	1	2	3	4	5
Component	% of Mass				
Bis-phenol A based epoxy resin	6.0	6.85	6.65	—	—
YD-134 - Kukdo	—	—	—	7.1	7.4
Liquid potassium glass	55.8	53.9	50.7	47.8	47.0
Diaminealkoxydifurfuryloxysilane	—	0.15	0.20	0.25	0.36
Furfuril alcohol	—	0.4	0.45	0.5	0.7
Formamide	8.1	7.9	7.8	7.6	7.5
Andezite powder	30.1	—	—	37.25	37.05
Diabase powder	—	30.8	34.2	—	—

Source: Reprinted from O. Figovsky and D. Beilin, "Adhesion Strength of Frecast Silicate Polymer Concrete Structure Member," *Journal Scientific Israel Technology Advanced* 13, no. 4 (2011): 96–102. With permission.

Surfactant oxyethyl cellulose was introduced in the adhesive composition to increase its homogeneity at a rate of 1% by weight of sodium silicate.

The results of the physical–mechanical testing and chemical resistance verification of the new adhesive composition are presented in Table 3.12.

Table 3.12 illustrates that the proposed adhesive composition is highly resistant to acidic environments at elevated temperatures. Reduction of epoxy resin of up to 6% leads to a decrease in adhesive strength up to 3.5 MPa, and an increase in its content of over 7.4% leads to a reduction in the chemical resistance coefficient of up to 0.81–0.83.

It may be noted that the introduction of the polymer additives increases cohesive and adhesive strength of the polymer silicate adhesive composition, keeping a high-acid resistance.

TABLE 3.12
Physical–Mechanical Properties and Coefficients of Chemical Resistance of the New Adhesive Composition Presented in Table 3.11

Index	Unit	Composition 1	2	3	4	5
Ultimate adhesion strength	MPa					
• at 20°C		3.6	3.8	3.9	4.2	4.3
• at 70°C		3.5	3.9	4.0	4.3	4.3
• at 120°C		3.5	3.8	4.0	4.1	4.2
Ultimate compression strength		48.1	50.5	52.0	53.5	55.0
Ultimate bending strength		17.6	18.8	19.8	19.7	20.4
Toughness	kJ/m^2	4.5	4.7	4.9	5.2	5.2
Coefficient of thermal expansion	1/°C	1.8	2.1	2.0	1.9	1.9
Coefficient of chemical resistance after 180 days exposure at 100°C in environments:						
• 40% H_2SO_4		0.64	0.94	0.95	0.95	0.97
• 60% H_3PO_4		0.62	0.95	0.94	0.98	0.96
• 10% CH_3COOH		0.67	0.92	0.91	0.90	0.90

Source: Reprinted from O. Figovsky and D. Beilin, "Adhesion Strength of Precast Silicate Polymer Concrete Structure Member," *Journal Scientific Israel Technology Advanced* 13, no. 4 (2011): 96–102. With permission.

Adhesive composition no. 5 is optimal from the standpoint of chemical resistance, and adhesive strength was tested on diffusion permeability by sulfuric acid.

For this purpose, two cubic samples of SPS concrete measuring 15 cm were glued. An end-to-end channel was created in the upper sample when forming. A glue joint was limited previously by strips from a sheet of polybutadiene. After curing of the adhesive compound, the strips were removed and the channel was filled with 10% sulfuric acid solution (Figure 3.4).

Observation of the samples showed no traces of sulfuric acid seepage through the glue.

Experimental Study and Discussion of Results

The strength of the adhesive composition was studied experimentally with SPC beams with continuous reinforcement rods. Three series of SPC beams were tested: a series of monolithic beams (I) and two series (II and III) with an adhesive joint in the middle of the span. Each series consisted of three samples.

The beams were produced from SPC concrete of optimal composition [5]:

Liquid glass (1.4 g/cm^3)	11.23%
Na_2SiF_6.(2700 cm^2/g)	1.68%
Diabase flour (2400 cm^2/g)	20.06%

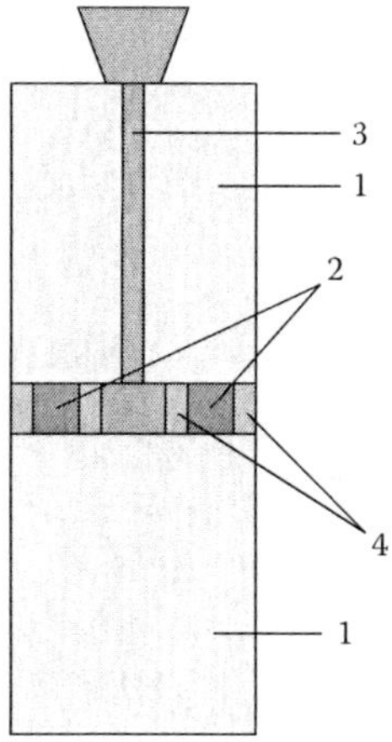

FIGURE 3.4 Aggressive environment testing of the adhesive joint: (1) SPS concrete samples, (2) adhesive composition, (3) channel filled by 10% solution of sulfuric acid, (4) sealant strips from chemical resistant vulcanized rubber. (Reprinted from O. Figovsky and D. Beilin, "Optimal Composition, Strength, and Chemical Resistance of Silicate Polymer Concrete," *Proceedings of ICPIC 2010: 13th International Congress Polymers in Concrete*, February 10–12, 2010, Madeira, Portugal. With permission.)

Quartz sand	26.71%
Granite chippings (5–10 mm)	30.98%
Tetrafurfuryloxisilane (TFS)	0.34%

Experimental samples of SPC beams are shown in Figure 3.5.

Beams of series I were tested to determine the load-bearing capacity on the bending moment. Beams of series II were tested to determine resistance to action of the shear forces. Both series were controls. Series III of the beams were tested to determinate the load-bearing capacity of adhesive joints of SPC adhesive joint structures. Epoxy–liquid glass adhesive (composition no. 5, Table 3.11) was accepted for joint grouting.

Beams of series III(A) and III(B) consisted of two parts. An overlap connection of reinforcement rods in the form of a closed loop was immersed in the adhesive joint; the length of the overlap is equal to 12 diameters of the working reinforcement. Such joints are widely used in plate structures.

Working reinforcement in beams of series III(B) is welded and then covered by the adhesive composition. Such joints are used in structures operated at high temperatures.

The physical–mechanical properties of the SPC, reinforcement, and adhesive as beam material are given in Table 3.13.

All beams were tested in simple bending; the load was applied in thirds of the span length $l = 1200$ mm, and the distance between loads was $a = 400$ mm (Figure 3.6).

Design load-bearing capacity of the beams is determined by condition [2]:

$$M \leq R_{cc}\, bx\,(h_0 - 0.5x) + R_{cr}\, A_{cr}(h_o - a) \tag{3.5}$$

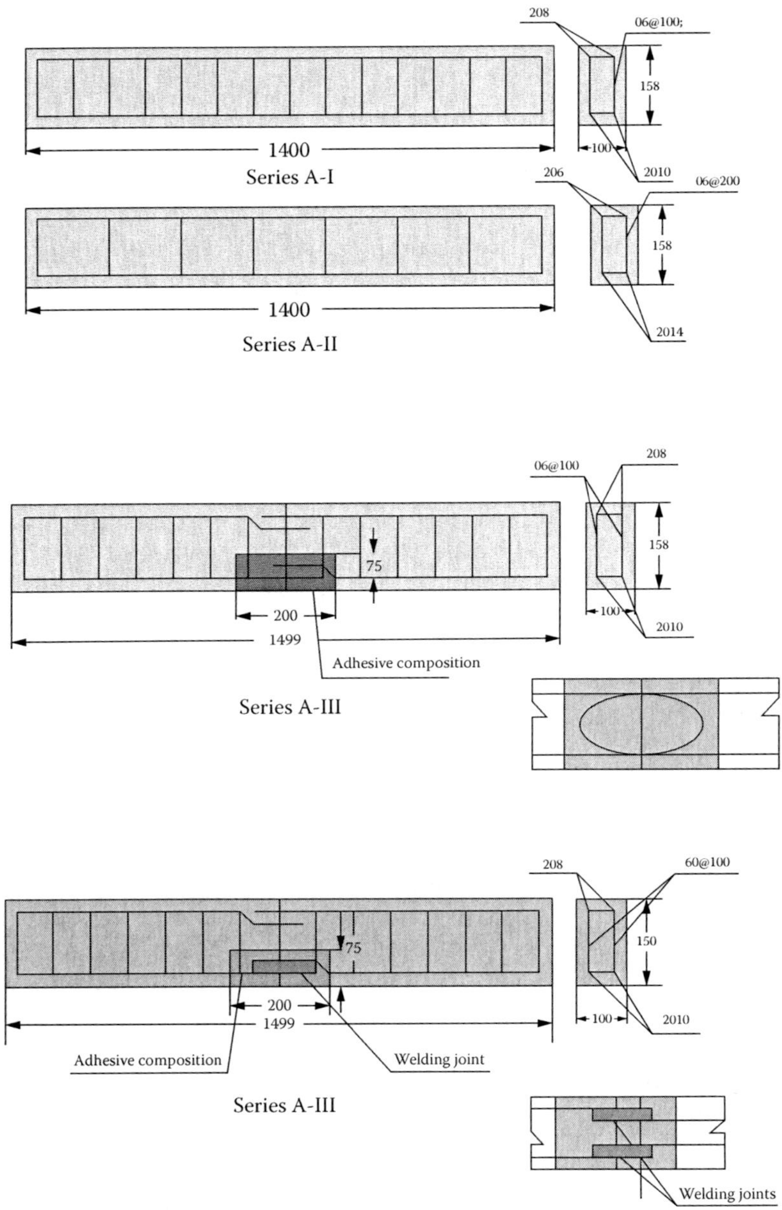

FIGURE 3.5 Experimental samples of SPC beams. (Reprinted from O. Figovsky and D. Beilin, "Optimal Composition, Strength, and Chemical Resistance of Silicate Polymer Concrete," *Proceedings of ICPIC 2010: 13th International Congress Polymers in Concrete*, February 10–12, 2010, Madeira, Portugal. With permission.)

TABLE 3.13
Physical–Mechanical Properties of the Beam Material

Index	SPC	Reinforcement	Adhesive Composition
Design strength at tension, MPa		$R_{cr} = R_{tr} = 210$	$R_a = 11.5$
Prism strength, MPa	$R_{cc} = 35$		
Module elasticity, MPa	$E_c = 2.4 * 10^4$	$E_r = 2.1 * 10^5$	$E_a = 1.0 * 10^4$

Source: Reprinted from O. Figovsky and D. Beilin, "Adhesion Strength of Precast Silicate Polymer Concrete Structure Member," *Journal Scientific Israel Technology Advanced* 13, no. 4 (2011): 96–102. With permission.

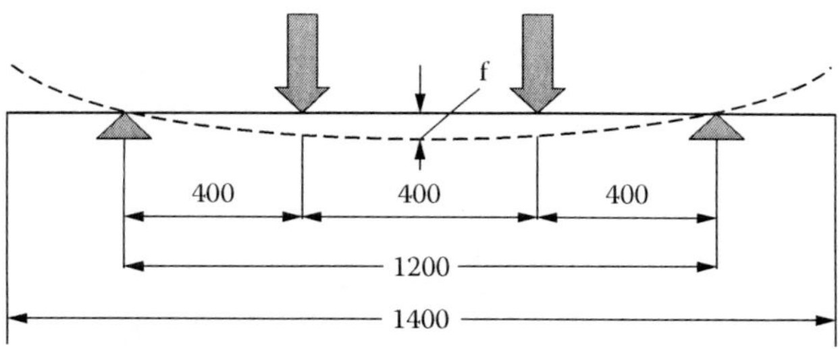

FIGURE 3.6 Scheme of SPC beams loading. (Reprinted from O. Figovsky and D. Beilin, "Optimal Composition, Strength, and Chemical Resistance of Silicate Polymer Concrete," *Proceedings of ICPIC 2010: 13th International Congress Polymers in Concrete*, February 10–12, 2010, Madeira, Portugal. With permission.)

where M is the design moment of the inner forces, R_{cc} is the prism strength of the SPC (Table 3.13), b is the width of the beam's section, h_0 is the working height of the beam's section, R_{cr} is the design strength of the compressed reinforcement, A_{cr} is the area of the compressed reinforcement, a is the thickness of the protective cover in the compressed zone of SPC; height of the compression zone of the of beam's section $x = (R_{tr}A_{tr} - R_{cr}A_{cr})/R_{cc}b$, R_{tr} is the design strength of the tensile reinforcement, and A_{tr} is the area of the tensile reinforcement.

The calculation of the strength of sections inclined to beam's longitudinal axis was performed by condition:

$$Q \leq Q_{fc} + Q_{fs} \tag{3.6}$$

where Q is the design shear force, Q_{fc} is the shear force experienced by the SPC, and Q_{fs} is the sum of the projections of the forces in the stirrups.

Destruction of the series AI beams began almost simultaneously in the tension and compression zones of SPC. Beams of series AII were destructed by inclined sections; the growth of the main inclined crack started in the SPC tension zone.

TABLE 3.14
Experimental and Calculated Values of Load-Bearing Capacity of SPC Beams

SPC Beams		Load-Bearing Capacity at Bending, kN		Strength of Inclined Section, kN		
Series	Sample Number	Experimental	Calculated	Experimental	Calculated	Kind of Fracture
I	1	10.0	7.53	—	—	Along compression and tensile zones
	2	11.4				
	3	10.8				
II	1	—	—	27.5	24.3	Along inclined section
	2			30.0		
	3			25.0		
III(A)	1	10.0		—	—	Along tensile zones
	2	9.0				
	3	8.5				
III(B)	1	10.2	7.53	—	—	Along compression zones
	2	8.7				
	3	9.2				

Source: Reprinted from O. Figovsky and D. Beilin, "Adhesion Strength of Precast Silicate Polymer Concrete Structure Member," *Journal Scientific Israel Technology Advanced* 13, no. 4 (2011): 96–102. With permission.

The destruction of the series AIII(A) beams began in the adhesive joint of the tension zone, while the destruction of the series AIII(B) beams started in the compression zone, which can be explained by *overreinforcement* of the lap-welded joint.

Experimental and calculated values of the load-bearing capacity of SPC beams are shown in Table 3.14.

In all cases, the experimental load-bearing capacity values of the samples was greater than the calculated average of 25%.

The mid-span deflections of the beams were measured during the experiment. The theoretical value of the deflection f is calculated by the formula:

$$f = Na\,(3l^2 - 4a^2)/24D, \tag{3.7}$$

where $D = E_c I_c$ is the stiffness of the SPC beam, $I_c = bh^3/12$ is the moment inertia of the beam transverse section, and h is the height of the beam's section.

Experimental and calculated deflections of the tested beams are given in Figures 3.7 and 3.8.

It is obvious that the experimental values of the deflections are less than calculated. In this case, the deflections of beams of control series I do not differ from series III beams.

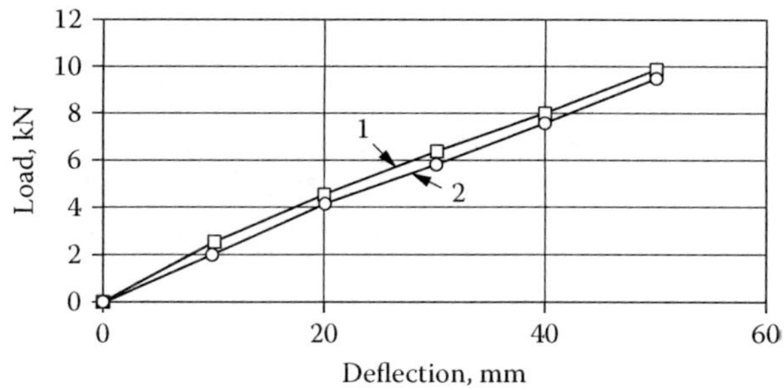

FIGURE 3.7 Relationship "load-deflection" for beams of series AIII: (1) calculated, (2) experimental. (Reprinted from O. Figovsky and D. Beilin, "Optimal Composition, Strength, and Chemical Resistance of Silicate Polymer Concrete," *Proceedings of ICPIC 2010: 13th International Congress Polymers in Concrete*, February 10–12, 2010, Madeira, Portugal. With permission.)

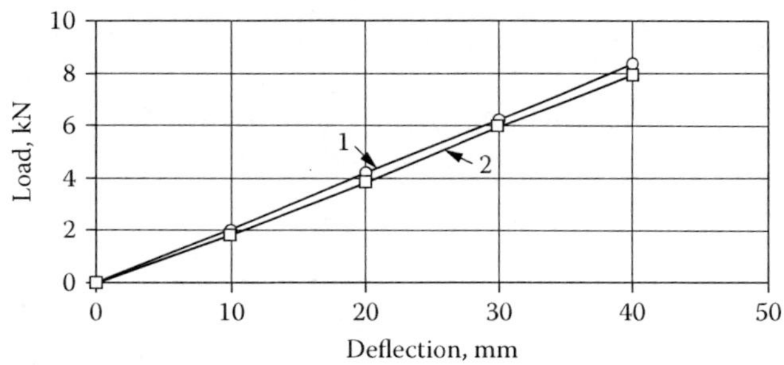

FIGURE 3.8 Relationship "load-deflection" for beams of series AI: (1) calculated, (2) experimental. (Reprinted from O. Figovsky and D. Beilin, "Optimal Composition, Strength, and Chemical Resistance of Silicate Polymer Concrete," *Proceedings of ICPIC 2010: 13th International Congress Polymers in Concrete*, February 10–12, 2010, Madeira, Portugal. With permission.)

Fracture and Crack Resistance of Silicate Polymer Concrete [6]

Crack Formation and Fracture of a Material in Terms of Fracture Mechanics

It is known that the real strength of a material is much less than the theoretical as a result of various defects in the form of pores, cracks, and so on. In response to loading in the vicinity of these defects, there are stresses that cause their growth, and thereby contribute to local or total destruction of the material [7].

A distinguishing characteristic of concrete, and in particular SPC, is a lack of soundness and equally strong bond between its structural elements. In addition, hardened concrete contains numerous micro and macro defects arising both during manufacture of the product (*technological cracks*) and in service (*operational cracks*).

Technological cracks are often caused by sedimentation processes associated with bleeding in the concrete mix and violation of integrity and uniformity of the hardened concrete structure. Sedimentation is directly connected with dependence of the velocity of ultrasonic waves in concrete on the wave front direction.

The liquid glass used as a binder in SPC contains 50%–60% water. With such a high water content of silicates solutions, the shrinkage strains arising during syneresis and removal of capillary moisture, lead to a reduction of the volume of material and the formation of micro- and macrocracks, especially at the binder–filler contact areas.

Crack formation in the manufacturing stage of SPC products and structures can be caused by stresses due to nonuniform distribution of moisture and temperature in the volume of concrete, and also as a consequence of unequal thermal expansion of SPC components during heat treatment. Loads associated with material handling, transportation, and erection can cause the formation of technological cracks.

This information allows us to conclude that technological cracks in SPC products and structures in the production stage are caused by general or local tensile strains in excess of the ultimate for freshly formed or hardened concrete. Shrinkage or temperature stresses, leading to internal cracking of a concrete, can occur not only in the manufacture of structures, but also in their operation process. Microcracks in concrete significantly worsen its operational properties and durability. Therefore, at an estimation of concrete quality is necessary to consider the origin and development of its microdefects and to define their influence on the behavior of the concrete in service [8].

In keeping with the theoretical concepts of fracture mechanics, the load-bearing capacity of a structure depends on the ability of a material to resist spreading of the main crack. The criteria for the beginning of a crack or fracture are established based on the limit equilibrium of a body with a crack [9,11].

It is assumed that the limit equilibrium state is reached if cracks develop and increase on the surface of the body volume under action of external loadings. In linear fracture mechanics, Irwin's force criterion and an equivalent Griffith's energy criterion completely determine the equilibrium condition of a continuum elastic body with a crack [9].

On the basis of the energy representations of the process of brittle fracture, Griffith stated that the work of external forces is spent for the volume deformation of a material and the formation of new surfaces—microcracks. Stresses in the top of microcracks of the loaded material are many times greater than the average stresses in a cross section of the sample. If the stress value at the top of the main crack is equal to theoretical strength, the crack quickly grows and the sample collapses.

From the energy point of view, this process is formulated as follows: "Failure occurs when the free energy attains a peak value at a critical crack length, beyond which the free energy decreases by increasing the crack length, i.e., by causing fracture" [10]. Using this procedure, Griffith found that

$$G = \sqrt{\frac{2E\gamma}{\pi}} \tag{3.8}$$

where E is the Young's modulus of the material and γ is the surface energy density of the material. Assuming $E = 62$ GPa and $\gamma = 1\ J/m^2$ gives excellent agreement with Griffith's predicted fracture stress with experimental results for glass [10].

Irwin proposed the concept of quasi-elastic fracture, which allows us to extend the limits of applicability of Griffith's theory [11]. Irwin's criterion is valid not only for brittle materials, but also for elastic-plastic materials with significant plastic deformation developing until the moment of actual destruction of a material.

The concept of quasi-brittle fracture implies that some materials that exhibit the characteristics of plastic materials at standard test, collapse in tests of the sample with a crack due to the quasi-brittle mechanism, that is, plastic deformation is concentrated in a narrow layer near the tip of a crack. For such materials, Griffith's criterion is relevant instead of the value of the surface energy to introduce the work of plastic deformation at the crack tip.

Irwin formulated a condition of crack propagation in a material, based on the notion of the asymptotic stress increasing in the volume element of the loaded body [11]: a crack starts to grow when the stress intensity factor K, associated with the applied stress σ and the crack length l by the ratio

$$K = \sigma\sqrt{\pi l} \tag{3.9}$$

reaches a critical value, which is constant for the given material under the given conditions (temperature, humidity, etc.). This is known as the *toughness* or *crack resistance* of a material.

Stress intensity factor K determines the nature and intensity of the stress state of a material in the vicinity of the crack tip. The ultimate value of stress intensity factor K_c defines the ability of a material to resist the formation and development of cracks and establishes the level of a material's stress state, the result of which is rapid and uncontrolled growth of cracks.

Thus, the crack begins to spread:

- If the energy release rate reaches a critical value (energy criterion of failure), or
- If the stress intensity factor reaches a critical value called the *fracture toughness* (force criterion).

According to linear fracture mechanics, fracture energy G_c of a linearly elastic body is given by

$$G_c = K^2_c/E \tag{3.10}$$

It should be emphasized that the concept of linear fracture mechanics can be applied only to materials with fine structure, such as a cement stone or fine-grained concrete, that is, to linear-elastic bodies. For such bodies, a condition of self-similarity for the prefracture zone exists

$$d << l, b, D \tag{3.11}$$

where d is the dimension of the end zone, l is the length of a crack, b is the linear size of the body in the direction of the crack, and D is the size of an average grain of the inclusion.

References [8,12] attempted to formulate the basic features of the behavior of cracks in a heterogeneous material such as concrete. The author [12] states:

1. The stress distribution in an inhomogeneous material even in the absence of cracks is significantly different from the distribution of stresses in a homogeneous body. This phenomenon is due to the different elastic properties of the components.
2. In the concrete, depending on the ratio of the properties of its components and characteristics of contact between these components, cracks can develop in different environments: in a cement matrix, in the filling and in the contact zone.
3. In an inhomogeneous material, cracks tend to move easily from one tougher than the hard stuff. Reverse is difficult, that is, possible, to stop cracks at the interface components.
4. A necessary and sufficient condition for the destruction of the sample material is the formation of one or more of the trunk, that is, through, cracks, causing division of the sample in two or more parts. Under this condition, the presence of a large number of samples, even non-through crack is not to say its destruction, on the other hand, education, for example, through the cracks in the longitudinal compression of the sample is considered equivalent to its destruction, even if formed part could still withstand compressive load.

Macrocracks due to brittle fracture in the top of a structural element is not, in general, a guarantee of the global destruction of the element. In brittle fracture, unstable crack growth begins immediately, but cracks can stay without breaking the design. This may be due to a low power-consumption design (not enough energy to provide dynamic crack growth) or a certain system of residual stress (cracked in the area of compression).

Thus, the reliability of the design in general is determined not only by the conditions that start the cracks, but also the kinetics of growth.

As mentioned previously, cracks start in the brittle fracture mechanism by the counter process, which includes the birth and development of cracks in the area of prefailure and its association with the macrocrack. Further development of the macrocrack, [13] further development of the microcrack is, perhaps, due to two alternative mechanisms.

The first mechanism is based on the idea that the growth of the macrocrack is due to the continuous nucleation of microcracks at its peak, which develop together with the macrocrack. In other words, the growth of a macrocrack is nothing more than a continuous nucleation of brittle fracture. Obviously, in the development of cracks by the first mechanism, a relatively large amount of continuous energy (as crack growth) must be provided with the necessary and sufficient conditions for the origin of macrofracture, which is associated with a lower or higher, but always with the presence of plastic deformation at the tip of the moving macrocrack.

The second possible mechanism of cracking is based on the following ideas. After the unification of microcracks with a macrocrack, continuous dynamic development of the macrocrack occurs without noticeable plastic deformation at the tip of rapidly developing cracks (not enough time to implement relaxation processes in the top).

The energy needed to start the crack is higher than the energy required for its development. Consequently, dynamic crack propagation in brittle fracture is most probably by the second mechanism.

Thus, the development of brittle fracture does not occur on a counter mechanism (as opposed to the start of brittle fracture), and is directly related to the growth of the main crack (macrocrack). By this means the possibility exists to use the concept of fracture mechanics, which reduces to the solution of the left-hand side of the formulas (3.9) and (3.10) (the parameters K, G, depending on the mode of loading design), and the right—the critical values that characterize the material's properties.

Experimental Research Methods and Machinery of Crack Resistance of Concrete

As a rule, crack resistance of coarse aggregate concrete is determined by nonequilibrium mechanical testing using samples with a cut that initiates a crack. Nonequilibrium tests are characterized by loss of the deformation process stability of the sample during the moment of strain localization at the maximum load, with corresponding dynamic development of the main crack [14]. However, such tests conflict with the condition in Equation (3.11), since in these concretes the size of the prefracture zone is 100–300 mm.

Nonequilibrium tests of concrete samples is defined by the energy accumulated in the sample at a certain stage of loading, which does not cause intense strain localization, and in so doing, the *ultimate equilibrium* is violated. A small increase in load results in disequilibrium between increasing external forces (including those due to the inertia of the testing machine) and sharply decreasing internal forces in concrete. An additional point to emphasize is that the use of linear fracture mechanics concepts, using the criteria in Equations (3.9) and (3.10), causes significant mistakes, because of the dynamic effect of nonequilibrium tests.

Thus, modern methods of nonequilibrium mechanical tests do not allow identification of deformation and fracture conditions of concrete samples, which leads to an ambiguous evaluation of structural material properties, in particular, the crack resistance of a concrete.

Equilibrium tests at the local deformation stage of a sample provide adequate changes of external forces to internal efforts of a material to resist with corresponding static development of the main crack. These tests are most appropriate when using bending or tension of the samples, because the fracture process will be defined by development of a unique *breakaway* type of crack, which allows determination of the actual surface area of the fracture. This means that the tests correctly provide the real physical processes of fracture of a concrete and the principles of nonlinear fracture mechanics with traditional mechanical characteristics of concrete and allows determination of a set of power and energy parameters of the material fracture.

Origin and Development of Cracks in Silicate Polymer Concrete

A schematic diagram of the device for the equilibrium test of concrete samples at a three-dot bend is shown in Figure 3.9.

A distinctive feature of the unit is the elastic steel ring providing stability and uniformity of the sample fracture by increasing the testing machine rigidity and the

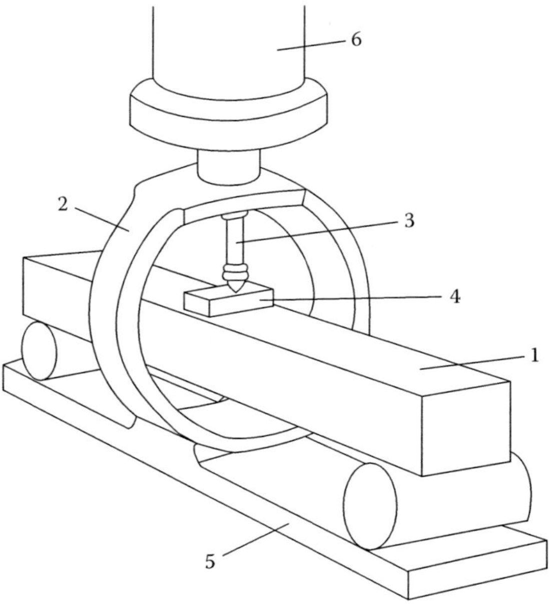

FIGURE 3.9 Test machine for equilibrium testing. (1) Test sample with initiated cut, (2) elastic steel ring, (3) punch, (4) distributive patch, (5) base, (6) press. (Reprinted from O. Figovsky, D. Beilin, and Yu. Zemlyanushnov, "Fracture and Crack Resistance of Silicate Polymer Concrete," *Journal Scientific Israel Technology Advanced* 14, no. 4 (2012): 38–48. With Permission.)

perception of excess elastic energy, which is released after the localization of deformation in the sample.

A block diagram of the control algorithm of equilibrium tests is shown in Figure 3.10. The compositions of test SPC samples are illustrated in Table 3.15.

Strain curves of the SPC samples for different types of monomer additives are shown in Figure 3.11.

$$W_c = \int_{f0_0}^{fmax} F\,df \tag{3.12}$$

The area under the curve of deformation in the coordinates: external load F is the deflection f with consideration for weight of the sample is a full energy consumption to fracture of the SPC sample W_c.

The work of fracture of an SPC sample can be presented by the sum of work spent on

- The development of microcracks = W_m
- Elastic deformation = W_e
- Local deformation = W_l

$$W_c = W_m + W_e + W_l \tag{3.13}$$

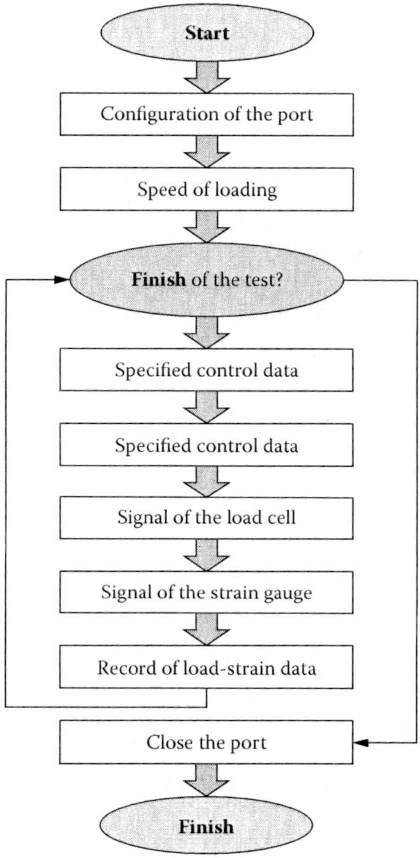

FIGURE 3.10 The control algorithm equilibrium tests. (Reprinted from O. Figovsky, D. Beilin, and Yu. Zemlyanushnov, "Fracture and Crack Resistance of Silicate Polymer Concrete," *Journal Scientific Israel Technology Advanced* 14, no. 4 (2012): 38–48. With Permission.)

In these conditions,

- Work spent for irreversible deformation (i.e., for fracture work) is

$$W_F = W_m + W_l \tag{3.14}$$

- Work spent for development of the main crack is

$$W_{mc} = W_e + W_l \tag{3.15}$$

The ratio of W_c to the area of a surface of fracture A is the specific fracture energy.

$$G_c = W_c/A \tag{3.16}$$

TABLE 3.15
Composition of SPC Samples

	Composition Number			
	1	2	3	4
Components	Content (% of mass)			
Liquid glass (1.4 g/cm^3)	11.23	11.23	11.23	11.23
Na_2SiF_6 (3200 cm^2/g)	1.68	1.68	1.68	1.68
Diabase flour (2400 cm^2/g)	20.06	20.06	20.06	20.06
Quartz sand	26.71	26.71	26.71	26.71
Granite chippings (5–10 mm)	40.32	39.98	39.98	39.98
Tetraetoxysilan (TEOS)	—	0.34	—	—
Furfuryl alcohol (FA)	—	—	0.34	—
Tetrafurfuryloxisilane (TFS)	—	—	—	0.34
Total	100.00	100.00	100.00	100.00

Source: Reprinted from O. Figovsky, D. Beilin, and Yu. Zemlyanushnov, "Fracture and Crack Resistance of Silicate Polymer Concrete," *Journal Scientific Israel Technology Advanced* 14, no. 4 (2012): 38–48. With Permission.

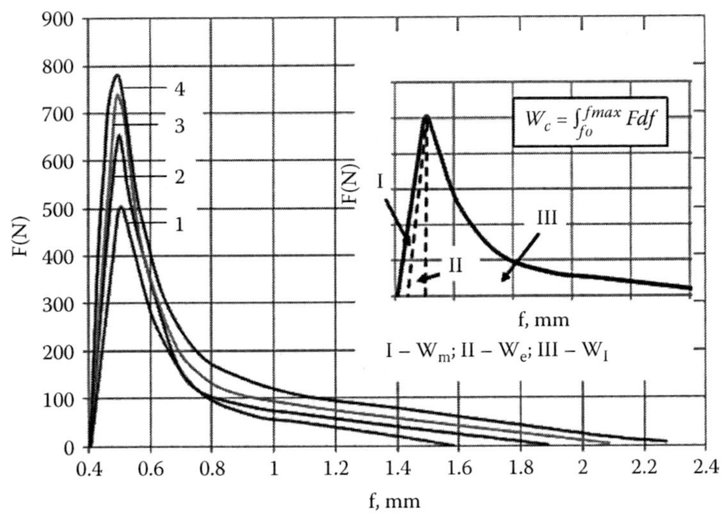

FIGURE 3.11 Strain curves of the SPC samples for different types of monomer additives. (1) Without additives, (2) with tetraetoxysilan (TEOS), (3) with furfuryl alcohol (FA), (4) with tetrafurfuryloxisilane (TFS). (Reprinted from O. Figovsky, D. Beilin, and Yu. Zemlyanushnov, "Fracture and Crack Resistance of Silicate Polymer Concrete," *Journal Scientific Israel Technology Advanced* 14, no. 4 (2012): 38–48. With Permission.)

Considering the heterogeneous structure of SPC, G_c is the integral value of the specific energy of fracture, which includes the surface energy due to irreversible deformations at the front of a crack during its development.

The force criterion of fracture (critical stress intensity factor) is calculated by the known formula of fracture mechanics:

$$K_c = \sqrt{G_c E} \tag{3.17}$$

The values of the energy and force fracture characteristics of silicate polymer samples, calculated according to the formulas (3.11)–(3.16) are presented in Table 3.16.

The analysis of the experimental results shows that the introduction of monomer additives leads to an increase in parameters W_c, G_c, and K_c. It is worth noting that introduction of the monomer additive TFS of 3% by weight of sodium silicate (composition 4) gives the best crack resistance performance.

As shown in [1, 2], monomeric additives are favorable for increasing the crack resistance characteristics of SPC due to

- Disaggregation of the system
- Destruction of aggregates by associated molecules
- Formation of a thixotropic material structure

Together, these lead to the formation of a homogeneous ordered structure in hardened concrete, a reduction of defects in the silica binder network, and thus to a decrease of internal stresses.

TABLE 3.16
Energy and Force Fracture Characteristics of SPC Samples

Strain Curve (No)	W_c*10⁻² (Nm)	W_m*10⁻² (Nm) / W_m/W_c	W_e*10⁻² (Nm) / W_e/W_c	W_l*10⁻² (Nm) / W_l/W_c	W_{mc}*10⁻² (Nm) / W_{mc}/W_c	W_F*10⁻² (Nm) / W_F/W_c	G_c (N/m)	(MN/m^{3/2}) K_c
1	17.17	0.58	3.06	13.53	16.59	14.11	90.12	1.25
		0.034	3.18	0.79	0.97	0.82		
2	17.52	0.42	4.09	13.01	17.10	13.43	93.95	1.53
		0.024	0.23	0.74	0.98	0.76		
3	21.89	0.32	3.75	17.82	21.57	18.14	122.0	1.84
		0.015	0.17	0.81	0.98	0.83		
4	27.12	1.96	3.65	21.51	25.16	23.47	147.3	2.16
		0.072	0.13	0.79	0.93	0.87		

Source: Reprinted from O. Figovsky, D. Beilin, and Yu. Zemlyanushnov, "Fracture and Crack Resistance of Silicate Polymer Concrete," *Journal Scientific Israel Technology Advanced* 14, no. 4 (2012): 38–48. With Permission.

Introduction of the monomer additives of orthosilicic acid esters into the SPC mix (compositions 2 and 4) contributes to formation of stronger contacts between the globules, reduction of porosity and shrinkage strains, improvement of silicate binder quality, and adhesion between the binder and coarse aggregate.

Analysis of the energy spent for fracture of SPC samples indicates that the relative work of micro–W_m/W_c and macro W_{mc}/W_c crack formation in the compositions with modifying additives leads to an increase in the crack resistance of these compositions. It is obvious that fluctuation of W_c, G_c, and K_c parameters depends on a change in relative values of the energy spent for elastic W_e/W_c and local W_l/W_c deformations.

Thus, the process of increasing the strength and crack resistance of silica polymer concrete after introduction of monomer additives to its composition is determined by the increase of the energy spent for the development of microcracks, elastic deformation, and the resistance of concrete to local deformation.

Analysis of the experimental results is presented in Table 3.4 shows that the traditional nonequilibrium mechanical testing of concrete leads to significant underestimation of its working efficiency.

Notice that total fracture energy $W_c = 27.12$ Nm with the modification of the SPS composition by TFS additive. The area under the ascending branch of the deformation graph no. 4 (Figure 3.11) determines the energy of microcrack development and elastic deformation as $W_m + W_e = 5.61$ Nm. However, in nonequilibrium tests, the energy of the local deformation $W_l = 21.51$ Nm is 80% of the total fracture energy W_c not considered.

The knowledge of the total fracture energy of concrete determined by equilibrium tests is especially important for structures operating in corrosive environments, shock, and cyclic mechanical loads because in this case the material acts, as a rule, beyond ultimate strength (descending branch of the deformation graph, Figure 3.11).

Therefore, the durability of the material is determined by its resistance to local deformation, that is, development of the main cracks.

Summary

- A small change of the quantity of liquid glass drastically changes the technical characteristics of SPC mixes.
- Introduction of TFS increases the harshness of SPC mixes.
- Strength and density of SPC compositions will increase with reduction of the liquid glass content.
- Optimal composition of SPC includes 11.23% liquid glass and 0.34% monomeric additives (FA or TFS).
- An SPC mix modified with TFS has higher strength indexes (cube strength up to 41MPa).
- Introduction of monomeric additives leads to significant reduction of shrinkage. Use of 3% TFS (from weight of liquid glass) reduces shrinkable deformations up to 0.06%.
- SPC samples with monomeric additives withstand the pressure of water to 0.6 MPa. The additives TFS and FA liquidate the filtering defects in the form of microcracks and conducting channels on a border gel filler.

- Corrosion of silica by hydrofluoric acid occurs rather quickly and leads to loosening of SPC composition and increasing of diffusive penetration. Because of this, application of the consolidating additive FA is desirable at pH ≤ 5.
- Experimental research has shown high mechanical characteristics of SPC structures with adhesive joints.

REFERENCES

1. Figovsky, O., and Beilin, D., "Improvement of Strength and Chemical Resistance of Silicate Polymer Concrete," *International Journal of Concrete Structures and Materials* 3, no. 2 (2009): 97–101.
2. Figovsky, O., and Beilin, D., "Optimal Composition, Strength, and Chemical Resistance of Silicate Polymer Concrete," *Proceedings of ICPIC 2010: 13th International Congress Polymers in Concrete*, February 10–12, 2010, Madeira, Portugal.
3. Figovsky, O., "Advanced Chemical Resistant Adhesives for Special Structures, *Kleben 7, Fachseminar Leistungsfähungkeit der Modernen Klebtechnik*, 11–13 Mai, *Interkantonales Technikum Rapperswil* (ITR) am Zûriuchsee, 1993 (in German).
4. Rudakov, O., Perzsev, V., Barsukova, L., and Usachev, S., "Efficiency of Application of Ethers of o-Silicic Acid for Modifying of Structure of Building Composites," *Materials of International Congress, Science and Innovation in Construction,* SIB-2008, Vol. 1, Book 2,Voronezh, 2008, 449–452 (in Russian).
5. Figovsky, O., and Beilin, D., "Adhesion Strength of Precast Silicate Polymer Concrete Structure Members," *Journal Scientific Israel Technology Advanced* 13, no. 4 (2011): 96–102.
6. Figovsky, O., Beilin, D, and Zemlyanushnov, Yu., "Fracture and Crack Resistance of Silicate Polymer Concrete," *Journal Scientific Israel Technology Advanced* 14, no. 4 (2012): 38–48.
7. Barbakadze, V., Kozlov, V., Mikul'skii, V., and Nikolov I., *Durability of Building Structures and Constructions from Composite Materials, AaBalkema.* ISBN-10: 9054102497, 1995.
8. Karimov, I., "Mechanics of Crack Formation at Fracture of a Concrete" (literature review) http://www.masterbetonov.ru/content/view/528/239.
9. Anderson, T.L., *Fracture Mechanics: Fundamentals and Applications*, Boca Raton, FL: CRC Press, 1991.
10. "Fracture Mechanics," Wikipedia, last updated May 31, 2013, http://en.wikipedia.org/wiki/Fracture_mechanics.
11. Irwin, G.R., *Fracture Dynamics: Fracturing of Metals*, Cleveland, OH: American Society for Metals, 1948.
12. Zaitsev, Yu., *Modeling of Deformation and Strength of Concrete Methods of Fracture Mechanics.* Moscow: Stroiizdat, 1982 (in Russian).
13. Karzov, G. P., Margolin, B. Z., and Schvetsova, V. A., *Physical–Mechanical Modeling of Fracture Processes.* St. Petersburg, Russia: Politechnica, 1993 (in Russian).
14. Nonequilibrium and Equilibrium Tests of a Concrete. http://www.npp-geotek.ru/employee/articles/concret/newarticle1.php?sphrase_id = 93290&print = Y

4 Nonisocyanate Polyurethanes Based on Cyclic Carbonates

Polyurethanes (PUs) are among the most demanded polymers, are required by modern technologies, and the worldwide consumption of PU has steadily increased. Of PU production 80% now consists of resilient and rigid foams while 20% remains solid PU (elastomers, coatings, adhesives, etc.) [1].

Despite its steady consumption, conventional PU has inherent weaknesses based on its molecular composition. The physical and mechanical properties of PU are mainly dependent on van der Waals forces. However, the strength of these bonds is significantly lower in energy [1]. Therefore, PU does not satisfactorily withstand loads, especially at elevated temperatures.

Also, the involvement of toxic components, such as isocyanates, in the fabrication process makes PU production extremely toxic and dangerous. People exposed to isocyanates can develop a range of short-term health problems. More seriously, isocyanate exposure can lead to long-term asthma and dermatitis if individuals become sensitized. Sensitization is a condition in which the breathing or skin conditions can return with increasing severity on further exposures to the original sensitizing agent or to similar substances, even at very low exposures [2].

Sources that do not contain isocyanates have long been sought for PU production. Nonisocyanate polyurethanes (NIPUs) based on polycyclic carbonates and polyamines have been known for more than 50 years. Fundamentals for the practical application of NIPUs in coatings, sealants, adhesives, and so on were developed in detail by O. Figovsky in the 1970s and 1980s [3]. Recently some reviews dedicated to the synthesis of cyclic carbonates and NIPU were presented [4–6]. The advantages of NIPU are described in detail in these works.

NIPU networks are created by the reaction between polycyclic carbonate oligomers and aliphatic or cycloaliphatic polyamines with primary amino groups [4]. This forms a cross-linked polymer with β-hydroxyurethane groups of a different structure—polyhydroxyurethane polymer. Since NIPU is obtained without using highly toxic isocyanates, the process of synthesis is relatively safe for both humans and the environment in comparison to the production of conventional polyurethanes. Moreover, NIPU is not sensitive to moisture in the surrounding environment.

Models of the two β-hydroxyurethane fragments of polymer chains formed in the case of dysfunctional starting materials is shown in Scheme 4.1.

Hydroxyl groups formed at the β-carbon atom of the urethane moiety increase adhesion properties. Plurality of intra- and intermolecular hydrogen bonds

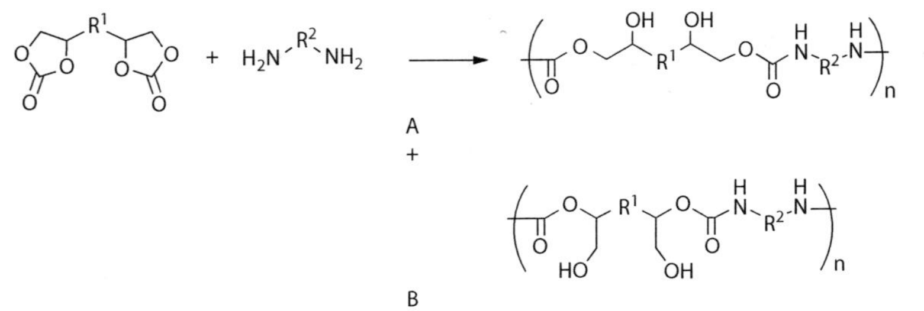

SCHEME 4.1 β-hydroxyurethane moieties of nonisocyanate polyurethanes: (A) with secondary hydroxyl group, (B) with primary hydroxyl groups. (Reprinted from O. Figovsky, L. Shapovalov, A. Leykin, O. Birukova, and R. Potesshnikova, "Nonisocyanate Polyurethanes Based on Cyclic Carbonates and Nanostructures Composites," *J. Scientific Israel Technology Advanced* 14, no. 4 (2012): 95–106. With permission.)

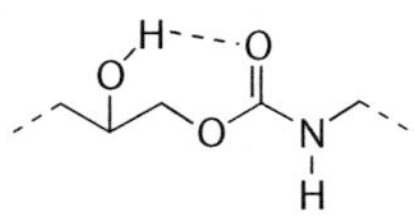

STRUCTURE 4.1 Hydrogen bonding in a hydroxyurethane. (Reprinted from E. Delebecq, J-P. Pascault, B. Boutevin, and F. Ganachaud, "On the Versatility of Urethane/Urea Bonds: Reversibility, Blocked Isocyanate, and Non-Isocyanate Polyurethane," *Chem. Rev.* 113 (2013): 80–118. With permission.)

(Structure 4.1) [7,8,42] as well as absence of unstable biuret and allophanate units [9] seem to be responsible for increased thermal stability and chemical resistance to nonpolar solvents. The reactivity of amine is shown in Scheme 4.2 [42].

The mechanism of the reaction of cyclic carbonates with amines, which creates nonisocyanate urethanes, was studied by means of quantum chemical calculation in terms of Density Functional Theory (DFT) by the PBE/TZ2P (Perdew, Burke, Ernzerhof gradient correlation functional using Triple Zeta Double Polarization basis) method using as examples the reactions of ethylene carbonate and propylene carbonate with methylamine [10,11]. Structural investigation revealed four cyclic isomers with an intramolecular hydrogen bond and six open conformers. This agrees well with previous results of infrared (IR) and nuclear magnetic resonance (NMR) spectroscopic investigations [4,7,8]. The reaction can proceed through the one- or multistage path involving one or two amine molecules. The second amine molecule acts as the catalyst of the process, resulting in a substantial decrease in the activation energy of the reaction.

Nonetheless, for the more than 50 years since the first publication in this field, NIPUs still do not have sufficiently broad application. This can be explained by certain features of these materials. Cyclic carbonate (CC) groups interact with aliphatic and cycloaliphatic polyamines at ambient temperatures more slowly than isocyanates with hydroxyl groups. The rate of this reaction is comparable to the rate of curing epoxy resins (ER) with amines. At the same time, the CCs react only with primary amino groups, in contrast to the ERs, which react with primary and with secondary amino groups. This results in a decrease in cross-linking density of the polymer network.

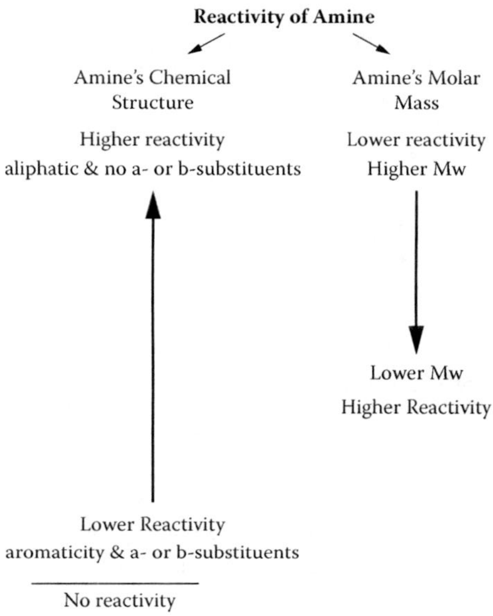

SCHEME 4.2 Scale of reactivity of various amines toward a model pentacyclocarbonate. (Reprinted from E. Delebecq, J-P. Pascault, B. Boutevin, and F. Ganachaud, "On the Versatility of Urethane/Urea Bonds: Reversibility, Blocked Isocyanate, and Non-Isocyanate Polyurethane," *Chem. Rev.* 113 (2013): 80–118. With permission.)

In this regard, preferential use of NIPU in hybrid systems based on copolymerization and modification of other polymer materials seems promising. Using an interpenetrating polymer network (IPN) principle in production of composite materials provides a unique possibility to regulate their both micro- and nanostructures and properties. By changing the IPN formation conditions (sequence of polymerization processes, ratio of components, temperature, pressure, catalyst content, introduction of filler, ionic group, etc.), it is possible to obtain a material with desirable properties.

POLYHYDROXYURETHANES AND HYBRID NONISOCYANATE POLYURETHANES

Methods of Polycyclic Carbonate Oligomer Synthesis

Five-membered cyclic carbonates are easily available as a result of the insertion of gaseous carbon dioxide into an oxirane ring (see review [12.]). Recent work in the field of new methods for preparing cyclic carbonates is dedicated primarily to the development of new catalytic systems and the synthesis of monofunctional compounds from epoxides and carbon dioxide (see, for example, reviews [13–16]). Monocyclic carbonates are used in a wide spectrum of applications: solvents, components of liquid electrolytes, reactive diluents, chemical intermediates, and so on. It should be noted that this preparation also solves the problem of chemical fixation and utilization of CO_2.

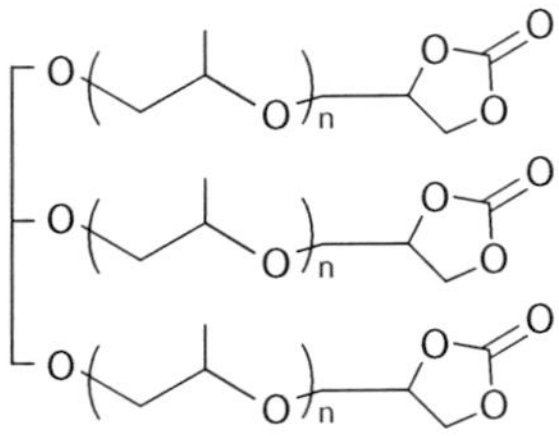

STRUCTURE 4.2 Example of polyfunctional cyclic carbonate–tricyclic carbonate on the base of propoxylated glycerin. (Reprinted from O. Figovsky and L. Shapovalov, "Cyclocarbonate-Based Polymers Including Non-Isocyanate Polyurethane Adhesives and Coatings," in *Encyclopedia of Surface and Colloid Science*, ed. P. Somasundaran, vol. 3, 1633–1653. New York: Taylor & Francis, 2006. With permission.)

At the same time there is the scarcity of commercially available polyfunctional cyclic carbonates (Structure 4.2). There are several methods for producing polymers or oligomers containing multiple cyclic carbonate groups:

- Direct catalytic CO_2 insertion into the appropriate polyoxirane compounds;
- Polymerization/copolymerization of monocyclic carbonates that also have unsaturated bonds;
- Interaction of monocyclic carbonates with second functionality and appropriate polyfunctional compounds.

A research team led by O. Figovsky synthesized aliphatic multifunctional cyclic carbonates from corresponding epoxies and carbon dioxide and NIPUs based on them [17,18]. The authors tested some compositions' polyfunctional carbonates synthesized in the laboratory, namely trimethylolpropane tricyclocarbonate (TMPTCC) and chlorine-contained aliphatic tricyclocarbonates (on the base of chlorine-contained aliphatic epoxy resins Oxilin™) and various diamines: 2-methylpentamethylene diamine (MPMD), Dytek® A, Invista Co.; meta-xylenediamine (MXDA), Mitsubishi Gas Chem. Co.; polyetheramineJeffamineEDR-148, Huntsman Co.; and diethylenetriamine (DETA), D.E.H.™ 20, Dow Chemical Co. The properties of these materials are shown in Table 4.1.

Some of the results are significantly better than previously achieved levels and offer good prospects for their practical use.

Multistage methods of synthesizing polyfunctional cyclic carbonates and hybrid nonisocyanate compositions using numerous epoxy–amine adducts are described in our work [19,20].

A special reactor for gas–liquid interaction is described by the authors in patent [18]. Due to the holes in the stirrer, CO_2 is involved intensively in the epoxy compound.

Applications of Unsaturated Cyclic Carbonate

Polymerization/copolymerization of unsaturated monocyclic carbonates is one of the methods of synthesizing polycyclic carbonates. The copolymerization behavior

TABLE 4.1
Properties of Polyhydroxy Urethanes

CC	Amine	Tensile Strength, MPa	Elongation,%	Water Absorption,%
TCCTMP	MPMD	33-47	3.3–3.8	n/a
CC Oxilin 5	MXDA	18	4	n/a
CC Oxilin 6B	EDR-148	0.8	10	4.0
CC Oxilin 6	DETA	1.6	16	N/A

Source: Reprinted from O. Figovsky, L. Shapovalov, A. Leykin, O. Birukova, and R. Potesshnikova, "Nonisocyanate Polyurethanes Based on Cyclic Carbonates and Nanostructures Composites," *J. Scientific Israel Technology Advanced* 14, no. 4 (2012): 95–106. With permission.

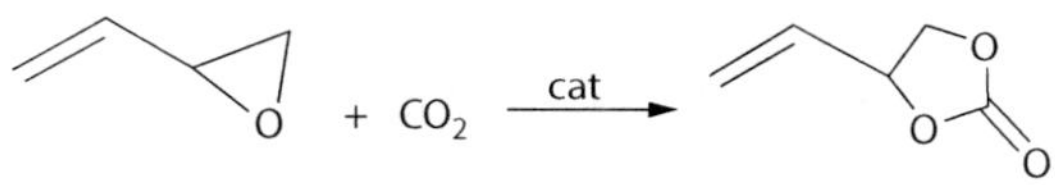

SCHEME 4.3 Synthesis of vinyl ethylene carbonate from 3,4-epoxy butene-1 and carbon dioxide.

of vinyl ethylene carbonate (VEC) was studied. VEC is readily prepared by the catalyzed addition of carbon dioxide to epoxy–butene (Scheme 4.3).

VEC is a sluggish monomer having free radical polymerization toward free radical polymerization. Due to the electronic structure of the double bond, VEC copolymerizes with vinyl ester monomers over a wide compositional range. However, VEC cannot be completely incorporated into an acrylic copolymer. When copolymerization with styrene is attempted, VEC is barely incorporated.

Copolymers with good solubility in common organic solvents and good compatibility with amine cross-linkers can be prepared with VEC levels up to 40%. Studies of the reaction rates of various amines with propylene carbonate in solution were used to identify amines that are suitable as cross-linkers. Hard and glossy clear coatings with good solvent resistance could be readily prepared, providing that the level of VEC in the copolymer was at an adequate level.

The monomers that have been explored most extensively are propylene carbonate methacrylate (PCMA) and propylene carbonate acrylate (PCA). These monomers are readily copolymerized with other commonly used unsaturated monomers to yield polymers with cyclic carbonate functionality. There are a few patents discussing the formation of coatings by the amine cross linking of these cyclocarbonate functional polymers. However, they do not appear commercially available. Thus, their use in the preparation of cyclic carbonate functional polymers has been limited.

Polyhydroxyurethane polymer usually has poor water resistance due to plurality of hydroxyl groups, but it is possible to prepare water-resistant materials in some formulations. For example, a cyclocarbonate acrylic was applied on a base of acrylic epoxy oligomer and a polymer was prepared with high water and weather stabilities

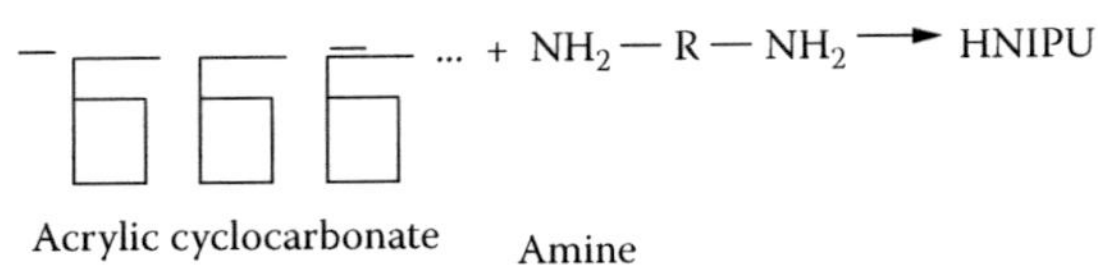

SCHEME 4.4 The process of acrylic NIPU curing.

by curing with primary amines (Scheme 4.4). Paint was developed with a curing temperature of 110°C in 2–3 hours.

The coatings produced from nonisocyanate polyurethanes have excellent water-resistance properties that are equivalent to the conventional polyurethanes and epoxy coatings combined [5,22,23]. The chemical resistance of NIPU materials can be increased by adding specific inorganic powdered substances during the fabrication process. These substances interact selectively with water and the aggressive medium (i.e., acids, alkalis, and salts), transforming the system of high-strength hydrate complexes into durable inorganic adhesive cements, thereby eliminating the defects of the material and increasing its strength.

Application of adhesives and sealants based on hybrid NIPU (HNIPU) for pasting metal surfaces is important in various industries [24]. Various compositions of adhesives and sealants were developed based on research results. Their physical-chemical properties are given in Table 4.2.

A number of compositions based on oligomer systems was developed [5]. These compositions contain hydroxy–amine adducts on a base of aliphatic mono- and polycyclic carbonates (Cycloate A™) as hardeners [25]. They are used for 100% solid floorings and have high abrasion resistance and good mechanical properties (for example F1 and F2 in Table 4.3).

One of the most effective ways to optimize these properties is the creation of hybrid polymeric materials that combine properties of high-molecular compounds of various classes with complementary properties, such as polyepoxides and polyurethanes. At the same time, it is preferable to avoid the use of raw materials with high toxicity, adverse effects on the environment, and so on, which consist of isocyanates—the main component of traditional polyurethanes [5]. One solution to this problem is to obtain polyepoxides-polyhydrourethanes based on epoxy compounds, amines, and cyclocarbonates without toxic isocyanates as raw materials. The presence of active hydroxyl groups contained in these compounds should give the effect of accelerating of curing process of epoxies with amines. Hydroxy-urethanes (also known as hydroxyalkyl urethanes or hydroxyalkyl carbamates) are formed by the reaction of cyclic carbonate groups (mainly five-membered groups) with primary amino groups.

A novel concept for generating new multifunctional modifiers was developed [26]. Hydroxyurethane modifiers (HUMs), which possess a wide range of hydrogen bonds, are embedded in an epoxy polymer network without a direct chemical interaction. Some new hybrid materials obtained by introducing hydroxyalkyl urethane into the polymer network structure, without any additional chemical reactions, are described in the following sections.

TABLE 4.2
Properties of Sealants and Adhesive Polymers Based on NIPU

Samples		Time of Curing @ 25°C, min	Adhesion to Al, MPa	Tensile Strength, MPa	Elongation, %	Hardness (ShoreA)	Weight Gain, 24 hrs @25°C (%)		
								`20%	
							Water	H_2SO_4	NaOH
Sealants	0-1758	225	14.0	23.0	2.5	96	1.7	7.8	0.6
	0-1817	80	15.0	8.0	38.0	86	7.7	14	2.4
	0-1818	190	16.0	18.0	5.5	94	3.7	9.2	1.9
	0-1823	100	12.0	38.0	5.6	94	4.9	9.8	1.2
	0-1824	60	9.0	22.0	6.3	93	4.7	8.3	1.6
	0-1828	90	9.0	36.0	3.0	94	1.7	6.2	0.5
Adhesive	0-1814	10	10.0	10.5	43	96	9	16	0.9
	0-1849	15	8	5	62	93	7	12	1.1
	0-1822	40	7.0	30.0	3.8	94	4.4	8.4	1.5
	0-1842	15	9.5	3.5	51.0	87	13.9	20	3.9
	0-1843	10	7.0	5.5	25.0	87	12.3	18	3.8
	0-1852	10	10	14	22	93	8	13	2.6

Source: Reprinted from O. Figovsky and L. Shapovalov, "Hybrid Nonisocyanate Polyurethane Adhesives," *Proceedings of the International Conference Polymer Bonding 2004,* April 27–28, 2004, Munich, Germany, 99–103. With permission.

TABLE 4.3
Composition and Properties of HNIPU Compounds

	Parts by Weight	
Composition	**F1**	**F2**
Hydroxyl-amine adduct "1" (on the base of Cycloate A™)	50.0	—
Hydroxyl-amine adduct "2" (on the base of Cycloate A™)	—	50.0
Epoxy resin D.E.R.® 324 of Dow chemical	45.0	40.0
Polycyclic carbonate Cycloate A™	5.0	—
Reactive acrylic oligomers (mixture) of Sartomer	—	10
Titanium dioxide	5.0	5.0
Carbon black	—	0.1
BYK®-A530 (surface active additive of BYK Co.)	2.0	—
BYK®-320 (surface active additive of BYK Co.)	—	1.5
	Values	
Properties	**F1**	**F2**
Mixed viscosity, 25°C, mPa·s	1,450	970
Pot-Life, 25°C, min	30–60	30–60
Tack free, 25°C, h	4	6
After 7 days at room temperature, substrate = concrete		
60° Film Gloss	100–105	115–120
Hardness, Shore D	70–80	70–80
Tensile strength, MPa	50–60	60–70
Elongation at break,%	5–7	3–4
Taber abrasion, 1000 cycles/1000 g, CS-17 wheel, mg	27	29
Impact resistance, N·m, ≥	20	20

HYDROXYURETHANE MODIFIERS

In many cases the use of epoxy materials in so-called field conditions (for industrial, construction sites, etc.) demand an increase in the reaction velocity, which is usually achieved by adding accelerators. At present, the widely used accelerators include alkyl-substituted phenols, benzyl alcohol, carboxylic acids (in particular, salicylic acid), and others. A major disadvantage of these accelerators is their tendency to migrate from the cured epoxy matrix during the exploitation, which could lead to a change in the physical properties of the polymer. They also act as a plasticizer of epoxy-based polymers,

and as a result reduce the polymer's chemical resistance. Thus, there is a need for new accelerator-modifiers that can provide faster curing of epoxy–amine compositions without negative side effects, and also improve the properties of the finished product.

Synthesis of Hydroxyalkyl Urethane Modifiers

The chemical reaction of obtaining hydroxyalkyl urethane modifier by reaction of monocyclic carbonate with polyamine is shown in Scheme 4.5 where R^1 is a residue

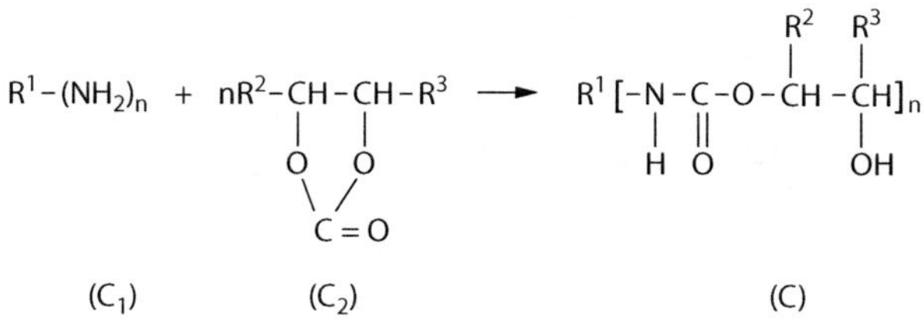

SCHEME 4.5 The process of HUM preparation. (Reprinted from O. Figovsky, O. Birukov, L. Shapovalov, and A. Leykin, "Hydroxyurethane Modifier as Effective Additive for Epoxy Matrix," *Scientific Israel Technological Advantages* 13, no. 4 (2011): 122–128. With permission.)

TABLE 4.4
Hydroxy-Alkyl-Urethane Modifiers

Name	CAS No.	Raw Materials	Calculated Molecular Weight	Viscosity 25°C. Pas
HUM-01	752243-46-2	Jeffsol PC, Vestamin TMD	362	79
HUM-04	119520-92-2	Jeffsol PC.IPD	375	27 (50°C)
HUM-D400		Jeffsol PC, Jeffamine D400	634	8
HUM-T403		Jeffsol PC, JeffamineT403	710	53

Source: Reprinted from O. Figovsky, O. Birukov, L. Shapovalov, and A. Leykin, "Hydroxyurethane Modifier as Effective Additive for Epoxy Matrix," *Journal Scientific Israel Technological Advantages* 13, no. 4 (2011): 122–128. With permission.

of the primary amine; R^2 and R^3 are the same or different and are selected from the group consisting of H, alkyl, hydroxyalkyl; and n satisfies the following condition: $n \geq 2$, C_1 = polyamine, C_2 = monocyclic carbonate, C = HUM.

The reaction was run at a stoichiometric ratio of cyclic carbonate and primary amine groups at ambient or elevated temperatures (up to 120°C). The process was controlled by infrared spectrometry until disappearance of the characteristic band at 1800 cm^{-1} ($\nu_{C=0}$) of cyclic carbonate.

It is possible to synthesize various hydroxyalkyl urethanes by changing the nature of the radicals R^1, R^2, and R^3.

Characteristics of the obtained modifiers are shown in Table 4.4. The properties of compositions based on various epoxy resins, hardeners, and modifiers are shown in Tables 4.5–4.8.

One can see a significant increase in the speed of the curing process, as well as a nontrivial increase in abrasion resistance, with a marked improvement in strength properties.

Dependences of curing characteristics and abrasion resistance from the content of HUM-01 are shown in Figures 4.1 and 4.2 [27].

TABLE 4.5
Properties of Compositions Based on Standard Epoxy Resin D.E.R. 331, Hardener Vestamin TMD, and Modifier HUM-01

Properties (23°C)	Value					
Content ofHUM-01,%	0	5	10	15	20	25
Pot life, min	90	60	44	38	27	25
Dry-to-touch, h	5	3,5	3	3	2	2
Hardness, Shore D	78	82	83	83	85	84
Tensile strength, MPa	44	50	50	49	58	50
Elongation at break, %	1.8	2.4	2.2	2.1	2.4	3
Abrasion resistance (weight loss mg/1000 cycles)	57	35	25	24	22	22

Source: Reprinted from O. Figovsky, O. Birukov, L. Shapovalov, and A. Leykin, "Hydroxyurethane Modifier as Effective Additive for Epoxy Matrix," *Journal Scientific Israel Technological Advantages* 13, no. 4 (2011): 122–128. With permission.

TABLE 4.6
Properties of Compositions Based on Hydrogenated Epoxy Resin ST-3000 (Kukdo, Korea), Hardener Vestamin TMD and Modifier HUM-04

Properties (23°C)	Value					
Content of HUM-1,%	0	5	10	15	20	25
Pot life, min	240	165	120	90	60	40
Dry-to-touch, h	8	5.5	4.5	4	3.5	3
Hardness, Shore D	80	81	81	81	82	78
Tensile strength, MPa	55	45	45	49	48	47
Abrasion resistance (weight loss mg/1000 cycles)	58	52	46	42	41	31

Source: Reprinted from O. Figovsky, O. Birukov, L. Shapovalov, and A. Leykin, "Hydroxyurethane Modifier as Effective Additive for Epoxy Matrix," *Journal Scientific Israel Technological Advantages* 13, no. 4 (2011): 122–128. With permission.

Thus, the wide range of hydrogen bonds of these modifiers allows them to be integrated into the epoxy–polymer network.

Hydroxyurethane Compounds from Renewable Plant-Based Raw Materials

Recently, the authors proposed a new method of producing a hybrid polyhydroxyurethane network composed of [28]:

(a) Reacting epoxidized unsaturated fatty acid triglycerides with carbon dioxide in the presence of a catalyst to obtain carbonated-epoxidized unsaturated fatty

TABLE 4.7
The Properties of Compositions Based on Resin DER 331, Hardener Vestamin TMD and Modifier HUM-D400

Properties (23°C)	Value					
Content of HUM-04,%	0	10	20	30	40	0
Pot life, min	90	85	75	70	65	90
Dry-to-touch, h	5	4.5	4	4	3.5	5
Hardness, Shore D	78	82	82	82	81	78
Tensile strength, MPa	44	62	62	68	57	44
Elongation at break,%	1.8	2.9	2.8	4	6.2	1.8
Abrasion resistance (weight loss mg/1000 cycles)	57	56	48	48	48	57

Source: Reprinted from O. Figovsky, O. Birukov, L. Shapovalov, and A. Leykin, "Hydroxyurethane Modifier as Effective Additive for Epoxy Matrix," *Journal Scientific Israel Technological Advantages* 13, no. 4 (2011): 122–128. With permission.

TABLE 4.8
Properties of Compositions Based on Various Epoxy Resins, Hardeners, and Modifiers

Properties (23°C)	Value		
Epoxy resin	D.E.R. 324	D.E.N. 431 Polypox R-14	Epalloy 5001Polypox R-11
Hardener	Ancamine 2432	Ancamine 1769	Ancamine 2379
Modifier	HUM-04, 10% HUM-D400, 10%	HUM-01, 15%	HUM-T403, 10%
Pot life, min.	23	28	21
Dry-to-touch, h	2.5	2	4
Hardness, Shore D	81	84	80
Tensile strength, MPa	4.2	5	5.7
Elongation at break, %	4.5	3.9	4.3
Abrasion resistance (weight loss mg/1000 cycles)	26	32	28

Source: Reprinted from O. Figovsky, O. Birukov, L. Shapovalov, and A. Leykin, "Hydroxyurethane Modifier as Effective Additive for Epoxy Matrix," *Journal Scientific Israel Technological Advantages* 13, no. 4 (2011): 122–128. With permission.

acid triglycerides, wherein conversion of oxyrane groups to 2-oxo-1,3-dioxolane groups (cyclic carbonate groups) for said carbonated-epoxidized unsaturated fatty acid triglycerides ranges from 35% to 85% (Scheme 4.6).

(b) Mixing and reacting the carbonated-epoxidized unsaturated fatty acid triglycerides with a compound having an amine functionality comprising at

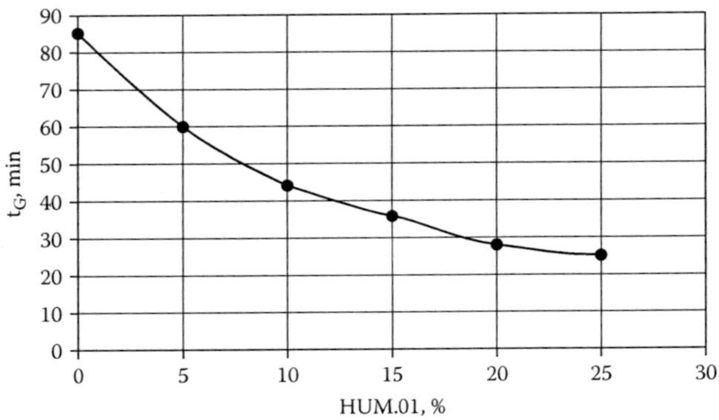

FIGURE 4.1 Influence of modifier HUM-01 (trimethyl-hexamethylene-diamine + propylene carbonate) on gel time of epoxy composition based on D.E.R. 331. (Reprinted from O. Figovsky, O. Birukov, L. Shapovalov, and A. Leykin, "Hydroxyurethane Modifier as Effective Additive for Epoxy Matrix," *Scientific Israel Technological Advantages* 13, no. 4 (2011): 122–128. With permission.)

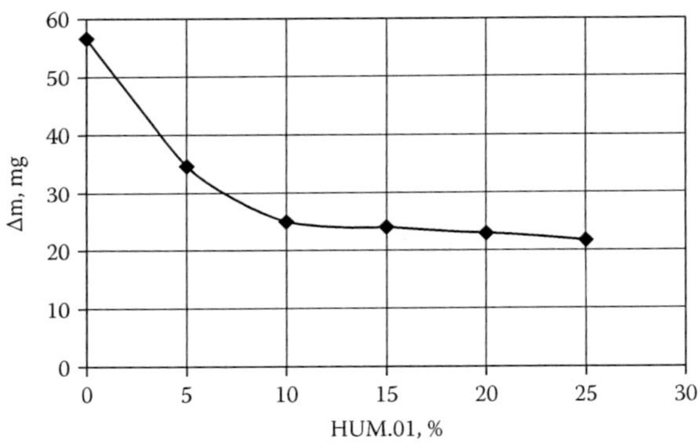

FIGURE 4.2 Influence the content of the modifier HUM-01 on the abrasion resistance of epoxy compositions (ASTM D4060, CS-17, 1 kg, 1000 cycles). (Reprinted from O. Figovsky, O. Birukov, L. Shapovalov, and A. Leykin, "Hydroxyurethane Modifier as Effective Additive for Epoxy Matrix," *Scientific Israel Technological Advantages* 13, no. 4 (2011): 122–128. With permission.)

least one primary amine group realized at stoichiometric or within nearly balanced stoichiometry.

(c) Mixing and reacting the product of (b) with a compound having amine functionality comprising at least two primary amine groups realized at excess of an amine-functional compound.

(d) Mixing the product of (c) with a compound having amino-reactive groups and selected from the group with:
- a compound having epoxy functionality,
- a mixture of the compound having epoxy functionality with carbonated-epoxidized unsaturated fatty acid triglycerides, a ratio of the sum of

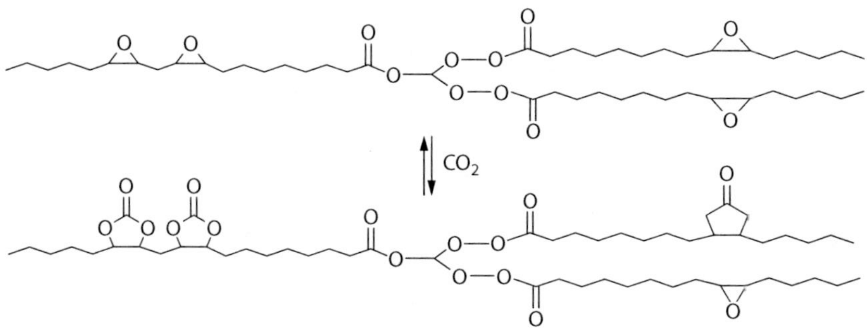

SCHEME 4.6 Partial carbonation of epoxidized vegetable oil. (Reprinted from O. Figovsky, L. Shapovalov, A. Leykin, O. Birukova, and R. Potesshnikova, "Nonisocyanate Polyurethanes Based on Cyclic Carbonates and Nanostructures Composites," *J. Scientific Israel Technology Advanced* 14, no. 4 (2012): 95–106. With permission.)

amino-reactive groups to the sum of amine groups being stoichiometric or within nearly balanced stoichiometry, and curing of the resulting composition at ambient temperature.

(e) The proposed method can significantly reduce time of synthesis and improve quality of the final products.

SILICON-CONTAINING AND NANOSTRUCTURED HYDROXYURETHANE COMPOUNDS

The concept of generating silica from alkoxysilanes by the sol-gel method within a macromolecular organic phase (in situ) is widely known. The organic and inorganic components of these materials are present as co-continuous phases of a few nanometers in lateral dimensions.

Aminosilanes are usually used as additives at a rate of 0.3%–2% in paints. The alkoxy groups are hydrolyzed by air or the surface humidity of a substrate. The newly built silanol-hydroxyl reacts with the surface hydroxyl groups and forms strong bonds. Organic groups, on the other hand, react with the organic components of the paint.

It is possible to hydrolyze aminosilane with water and, by doing so, prepare polyfunctional aminosilane (Schemes 4.7–4.9) [29].

Structures of hydrolyzed aminosilanes may be very different. Formation of nanoparticles is influenced by moisture. Nanoparticles are formed by addition of silica via the sol-gel process.

Amino-terminated oligomers are used for hardening cyclocarbonate-terminated oligomers [29–31]. Today, for thermostable epoxy compositions, aryl amines are used, which are toxic and require curing temperatures of more than 80°C. For thermostable epoxy and cyclocarbonate compositions, combinations of these oligomers are used with silicones [29].

The proposed dendro-aminosilane hardeners give the possibility for the introduction of siloxane fragments into the aromatic structure of bisphenol A (BPA)

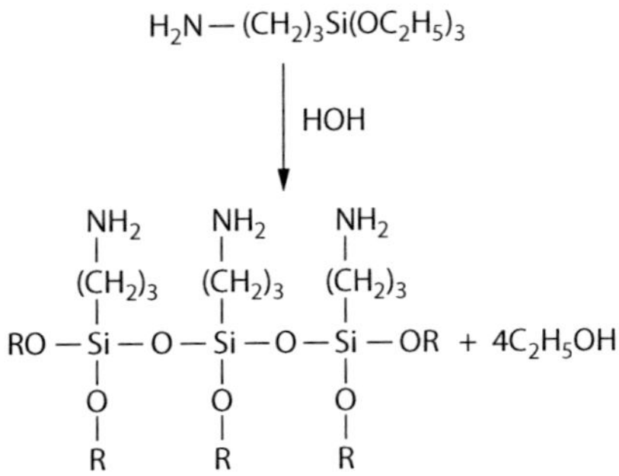

SCHEME 4.7 Hydrolysis–condensation processes for 3-aminopropyl triethoxysilane. (Reprinted from O. Figovsky and L. Shapovalov, "Cyclocarbonate-Based Polymers Including Non-Isocyanate Polyurethane Adhesives and Coatings," in *Encyclopedia of Surface and Colloid Science*, ed. P. Somasundaran, vol. 3, 1633–1653. New York: Taylor & Francis, 2006. With permission.)

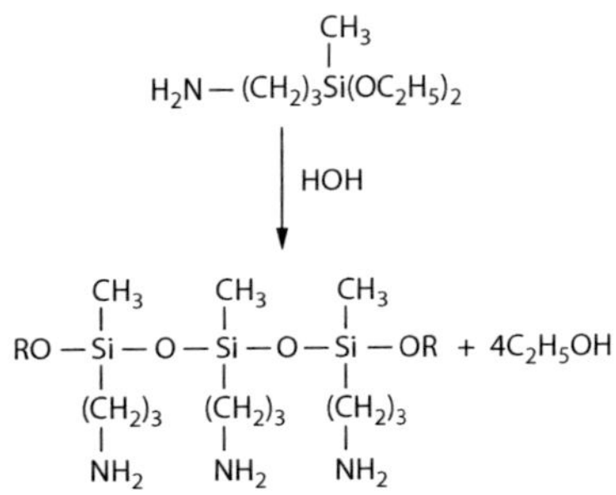

SCHEME 4.8 Hydrolysis–condensation processes for 3-aminopropyl diethoxymethylsilane. (Reprinted from O. Figovsky and L. Shapovalov, "Cyclocarbonate-Based Polymers Including Non-Isocyanate Polyurethane Adhesives and Coatings," in *Encyclopedia of Surface and Colloid Science*, ed. P. Somasundaran, vol. 3, 1633–1653. New York: Taylor & Francis, 2006. With permission.)

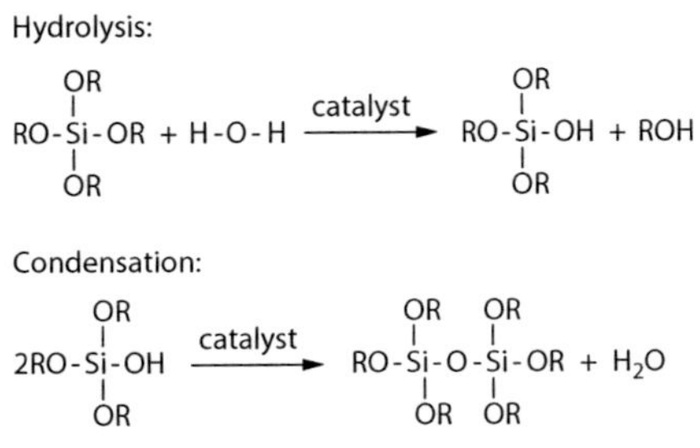

SCHEME 4.9 Hydrolysis–condensation processes for tetraalkoxysilane. (Reprinted from O. Figovsky and L. Shapovalov, "Cyclocarbonate-Based Polymers Including Non-Isocyanate Polyurethane Adhesives and Coatings," in *Encyclopedia of Surface and Colloid Science*, ed. P. Somasundaran, vol. 3, 1633–1653. New York: Taylor & Francis, 2006. With permission.)

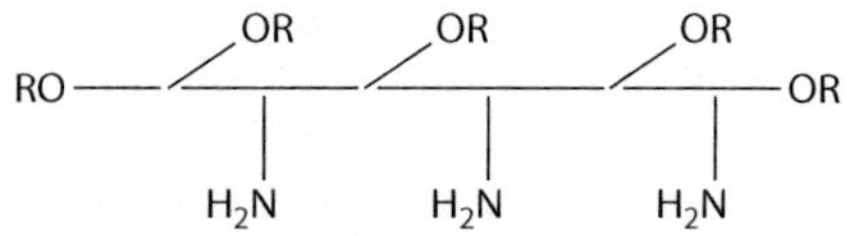

STRUCTURE 4.3 Schematic curing agent. (Reprinted from O. Figovsky and L. Shapovalov, "Cyclocarbonate-Based Polymers Including Non-Isocyanate Polyurethane Adhesives and Coatings," in *Encyclopedia of Surface and Colloid Science*, ed. P. Somasundaran, vol. 3, 1633–1653. New York: Taylor & Francis, 2006. With permission.)

epoxy–amine and cyclocarbonate network polymers, which improves the service properties of the network polymer.

The curing agent was prepared by hydrolysis of a mixture of aminosilane and organosilanes at room temperature with the evaporation of alcohol in a solution of alcohol (Structure 4.3).

Groups (OR) are silanol alkoxy groups, which react with the active centers of the substrate by reaction with –OH groups and this forms a stable bond. Additional hydrolysis of organosilane oligomers creates a secondary nanostructured network polymer.

A further development of silane curing agents are silanes in which one or two of the silicon functional groups are replaced by the inert methyl or aryl groups, which impart special effects (adhesion, mechanical properties, etc.).

Acrylic nonisocyanate polyurethane has high weather stability, but only if curing was at 110°C. A room temperature (RT) coating was prepared by using dendroaminosilanes. Cyclocarbonates on a base of acrylic resins were prepared by using the reaction of epoxy groups with CO2 in the presence of a catalyst. Synthesis of the curing agent was provided by hydrolysis of amino-propyl methyldiethoxysilane, tetraethyl orthosilicate (TEOS), and water. The gel fraction that occurred after curing for 24 hours at RT is 84%. Nanoparticles are formed by reaction of the alkoxy groups with water.

Novel nanostructured hybrid polymer compositions were synthesized on a base of epoxy functional components, cyclic carbonate components, amine-functional components, and acrylate (methacrylate) functional components, wherein at least one epoxy, amine, or acrylate (methacrylate) component contained alkoxysilane units [32]. The composition is highly curable at low temperatures (approximately 10°C to 30°C) and generates a nanostructure (Figure 4.3) under the influence of the forming of active, specific hydroxyl groups by reaction of cyclic carbonates with amine functionalities. These hydroxyurethane functionalities activate hydrolytic polycondensation of alkoxysilanes by means of atmospheric moisture, thus producing an organic–inorganic nanostructure without a special procedure of water embedding or addition of nanofillers. The cured composition has excellent strength–stress properties, adhesion to a variety of substrates, appearance, and resistance to weathering, abrasion, and solvents.

Sprayable Foam

A basic composition and a technique for mixing and foaming were developed for an insulating foam applied by spraying. The material is based on the use of the synthetic raw materials. A standardized procedure for obtaining polyurethane foams

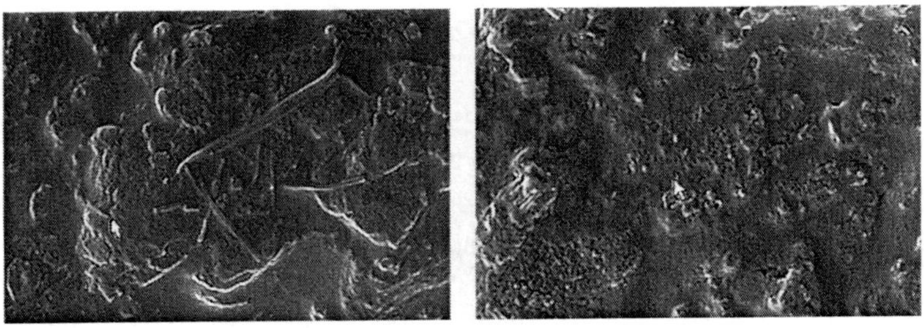

FIGURE 4.3 Images of the fractured surfaces of the cured compositions at 20,000x magnification (AURIGA CrossBeam): (A) epoxy-silane-amine composition, (B) neat epoxy-amine composition. (Reprinted from E. Delebecq, J-P. Pascault, B. Boutevin, and F. Ganachaud, "On the Versatility of Urethane/Urea Bonds: Reversibility, Blocked Isocyanate, and Non-Isocyanate Polyurethane," *Chem. Rev.* 113 (2013): 80–118. With permission.)

was approved. Technical specifications of this newly developed composition correspond to mean values of a similar urethane-based thermal insulation (the ultimate strength of the new material is about 1.5 times higher than that of the polyurethane foam, and the heat-insulating properties are at the same level as in polyurethane foam with open pores). Possibility of application by spraying was tested on pilot equipment.

The composition of the foam was studied by the use of renewable raw materials, including new HUM. As a result, hard and elastic foams were obtained with properties not inferior to polyurethane foams. The foams were produced in the laboratory only (Table 4.9).

UV-Curable HNIPU Floorings and Coatings

UV-curable concrete floor coatings provide a durable, high-performance, and eco-friendly solution. These thin-film coating systems cure instantly, thereby minimizing the downtime of any facility. Other benefits of these coatings include excellent chemical resistance, little odor, the ability to coat in cold conditions, and ease of cleanup [33].

The surface preparation and application of UV coatings is similar to that of traditional concrete floor coatings. UV coating systems include both clear and pigmented systems. The clear system consists of a primer and a topcoat that is available in different finishes ranging from high gloss to matte. The topcoat finish can be further enhanced by broadcasting additives for decorative or performance purposes.

Single-coat systems are also available. The thin film thickness can range between 0.3 and 0.8 mm.

Unlike conventional UV-curable coatings, formulations developed by our company contain up to 60% vegetable oil derivatives, such as modified soybean oil. This allows a composition of higher elasticity while maintaining the basic strength characteristics. The use of our compounds improves the adhesion of the cured composition to concrete, allows a reduction in the number of coating layers to two, and for special coatings even to a single layer. The uniqueness of this

TABLE 4.9
Rigid HNIPU Foam

Rigid Foam Insulation	Standard	Properties
	ASTM DT196	Viscosity (Brookfield RVDV II, Spindle 29, 20 rpm) at 25°C, cP
2800–3200		Base A
3600–4100		Base B
≤3700		A + B (3–5 sec after mixing)
8-10		Pot life at: 25°C (77°F), s
Compliant	ASTM D2369	VOC
2–4		Gel time, s
30–40		Touch dry, s
15–20		Curing for transportation, min
White		Appearance of rigid foam
0.2–0.4	ASTM D1621	Compressive Properties of Rigid Cellular Plastics,
24 hours, MPa		
30–40	ASTM DT1622	Apparent density of rigid cellular plastics, kg/m^3
4.5–5.0	C 518	Thermal transmission resistant (R-value) hr.ft^2 °F/Btu.in

Source: Reprinted from O. Figovsky, L. Shapovalov, A. Leykin, O. Birukova, and R. Potesshnikova, "Nonisocyanate Polyurethanes Based on Cyclic Carbonates and Nanostructures Composites," *Journal Scientific Israel Technology Advanced* 14, no. 4 (2012): 95–106. With permission.

compound is the possibility to apply one layer up to 0.8 mm thick, at the rate of polymerization, which allows the use of standard curing technology and standard equipment.

The introduction of our new HUM, based on vegetable raw materials and adducts obtained on their basis, allow improved hardness and wear resistance, while maintaining the other properties of the system (Table 4.10).

The uniqueness of the developed formulation and the possibility of coating concrete, sometimes without a primer, with a layer thickness of 0.3–0.8 mm allows the covering of even open areas. Application is done by spraying, which eliminates the negative effects of sunlight during the coating process, and uses sunlight during the curing process, thus reducing the total polymerization time even more.

Nonisocyanate Polyurethanes for Use in Monolithic Industrial Floor Coverings and Coatings

A novel nonisocyanate polyurethane binder for monolithic flooring and industrial paint compositions was produced on an industrial scale [34]. The two component binders have unique properties that combine the best mechanical properties of polyurethane and the chemical resistance of epoxy binders [35,36]. The polyurethane binders do not present health hazards because they do not consist of isocyanate

TABLE 4.10
Properties of HNIPU UV-Cured Flooring Compared to Conventional UV-Cured Flooring

Properties	Standards	Conventional UV-Cured Flooring	HNIPU UV-Cured Flooring
Adhesion	ASTM D 3359-07, B	3B	5B
Pencil hardness	ASTM D 3363-05	3H	4H
Solvent resistance	ASTM D5402-06	200+	200+
Gloss	ASTM D 523	84	90
Abrasion resistance, CS-17, 1000 grams, 1000 cycles, mg	ASTM D4060	150-200	100
Thickness applied, mm		0.065–0.1	0.3–0.8
Primer		Required	Not required

Source: Reprinted from O. Figovsky, L. Shapovalov, A. Leykin, O. Birukova, and R. Potesshnikova, "Nonisocyanate Polyurethanes Based on Cyclic Carbonates and Nanostructures Composites," *Journal Scientific Israel Technology Advanced* 14, no. 4 (2012): 95–106. With permission.

components at any stage of preparation. The environmentally friendly binders are insensitive to the moisture in the air or the coated surface and provide for marking monolithic nonporous materials with decreased permeability.

Industrial floors are one of the basic elements of a building and simultaneously its most loaded part. Floors experience the enormous loadings connected with abrasive and mechanical deterioration, thermal loadings, chemical and impact influences, and so on. The general requirements for industrial floors are [5]

- Wear resistance
- Mechanical load resistance
- Impact resistance
- Chemical resistance
- Temperature resistance
- Impenetrability for liquids
- Crack resistance
- Slip resistance
- Fire safety
- Effective sound absorption
- UV radiation resistance
- Antistatic, current-carrying or dielectric properties
- Attractive appearance and color palette
- Ease of cleaning and service
- Fast construction/application
- Placement of a floor on freshly placed or damp concrete
- Hygienic properties of coverings

The fields of application of monolithic floor coverings are quite extensive and highly diversified. These are industrial workshops and adjoining platforms, garage complexes and multistory parking, car-care centers and car washes, warehouse and shopping centers, industrial refrigerators and freezing chambers, sports constructions, corridors, staircases, and so on.

In the large variety of polymeric materials applied for flooring, polyurethane coverings have a special place. Polyurethane floors have important advantages over all known coverings (concrete, linoleum, tile etc.) on a lot of parameters. Monolithic coverings on a polyurethane basis have high mechanical strength at compression and tension, and wear resistance. They are elastic, have high chemical stability to aggressive environment action including acids, alkalis, solvents, oils, and so on. During operation, a floor covering on a polyurethane basis endures a wide range of temperatures and greater impact loadings.

NIPU materials could be applied for monolithic floor coverings having high chemical, crack, wear, and fire resistance and minimal absorption at toxic or radioactive contaminations. The coverings are very homogeneous and do not require any solvents during preparation, can be operated in high-humidity media, meet special sanitary and hygienic requirements, ands are ideally suited to industrial, institutional, and commercial uses.

The mechanical and chemical resistance properties of two-component nonisocyanate polyurethane flooring are illustrated in Tables 4.11 and 4.12.

Each test sample includes 150 weight parts (w.p.) of quartz filler per 100 w.p. of the binder.

The chemical resistance of nonisocyanate polyurethanes can be increased by adding inorganic powdered substances during the fabrication process. The substances

TABLE 4.11
Mechanical Properties of Two-Component Filled Nonisocyanate Polyurethane Covering

Type of Cyclocarbonate Oligomer	Functionality of Cyclocarbonate Oligomer	Tensile Strength/Elongation, (MPa/%)				
		Functionality of Compound				
		2.9t	3.5	4.3	5.1	2
I	2	1.9/710	2.7/565	4.9/325	5.3/108	—
I	3	8.8/220	13.7/114	15.3/90.5	15.4/63.5	26.6/88.5
I	5	21.3/72	21.8/50	18.7/31	17.2/22	20.3/48.5
II	2.45	8.2/315	17.3/98	23.9/47	2.6/19.6	30.8/20.5
II	3.1	19.3/88	22.7/62	23.8/37	18.2/18.8	33.9/19.2
II	3.9	19.9/82	26.1/42	25.4/28	17.1/14.3	34.2/18.3
II	3.9	16.9/73	13.7/30.5	12.7/24.5	10.6/13.6	27.1/16.9

Source: Reprinted from O. Figovsky and D. Beilin, "Building Materials Based on Advanced Polymer Matrix," *Journal Scientific Israel Advanced Technology* 10, no. 3 (2008): 1–119. With permission.

TABLE 4.12
Coefficients of Chemical Resistance of the NIPU Covering

Environment	Coefficients of Chemical Resistance
H_2SO_4 10% at the 60°C	0.90–0.95
NaOH, 10% at the 60°C	t0.95–1.0
H_2O, at the 60°C	0.95–0.90

Source: Reprinted from O. Figovsky and D. Beilin, "Building Materials Based on Advanced Polymer Matrix," *Journal Scientific Israel Advanced Technology* 10, no. 3 (2008): 1–119. With permission.

selectively interact with water and aggressive media (acids, alkalis, and salts) by forming a system of high-strength hydrate complexes in durable inorganic adhesive cements. At interaction with the medium, crystal hydrates penetrate into micropores and microcracks of a material. This process "heals" a material and increases its strength. During the process of crystal-hydrate formation, the volume and the specific contact surface of an active additive expand into the constant volume of the polymeric matrix, as a result of which, the adhesion on the polymer–crystal boundary becomes stronger.

The suggested monolithic floor covering can find application in the following venues:

- Enterprises of the chemical, paint-varnish, and pharmaceutical industries
- Enterprises of the electronics industry and precise mechanical engineering
- Enterprises of the machine-building and aerospace industries
- Enterprises of the wood-working industry
- Medical institutions and objects of consumer services
- Trading premises and warehouses
- Premises with increased decorative requirements: trading and showrooms, television studios, etc.

Colored glass chips can be inserted in the NIPU flooring with a subsequent coating of transparent protective varnish. Binder with rubber granules and quartz can be used on ramps and other critical places for increase of adhesion. The coating can be of any color. Figure 4.4 illustrates the recommended structure of the floor covering.

New High-Quality Monolithic Flooring

Monolithic two-component nonisocyanate floor covering ECPU 2851™ is placed on a base of cyclocarbonate and epoxy oligomers (Figure 4.5). The covering is destined for manufacture of seamless indoor floors for areas where both cleanliness and durability are crucial. It does not contain organic solvent, harmful or toxic substances; it possesses high chemical and abrasive resistance; it can be operated in high-humidity

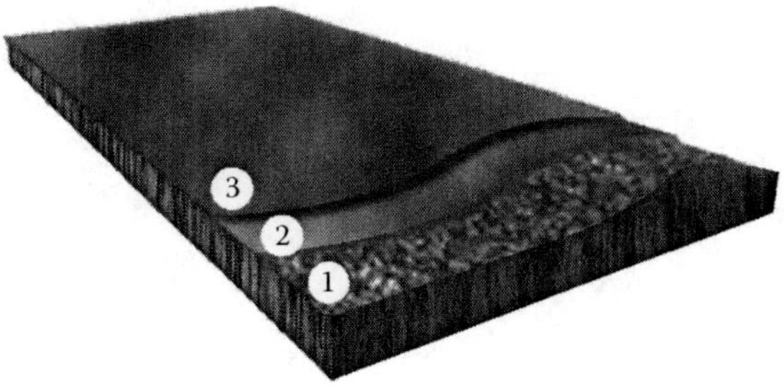

FIGURE 4.4 Structure of flooring: (1) substrate, (2) primer, (3) covering with thickness 0.5–3mm. (Reprinted from O. Figovsky and D. Beilin, "Building Materials Based on Advanced Polymer Matrix," *J. Scientific Israel Advanced Technology* 10, no. 3 (2008): 1–119. With permission.)

FIGURE 4.5 Floor covering ECPU 2851. (From Phazer Inc., http://www.phazerinc.com/phazerthane.html; and Nanotech Industries, http://nanotechindustriesinc.com/GPU.php.)

media; it meets special sanitary and hygienic requirements; and it is ideally suited to industrial, institutional, and commercial uses.

The properties of such hybrid nonisocyanate polyurethane coverings are given in Tables 4.13–4.16.

The company Polymate Ltd-INRC (Israel) has developed a number of new two-component flooring and coating compositions based on hybrid nonisocyanate polyurethanes using hydroxyurethane modifiers. Properties of the new materials are presented in Table 4.16 [5,37,38].

Testing Methods and Procedures for Industrial Floors [39]

Standards for finished industrial floors and their execution do not currently exist, and nor do standards for testing of the properties of the flooring materials. The current approach to the design and construction of industrial pavements in many cases is based on wide, mostly negative experiences over time.

Nowadays, with development of industrial and storage technology, industrial floors have to fulfill many special requirements for intended use, which are discussed in [39–41]. For this reason, regulations are necessary for proper design, construction, and control of industrial floors and related testing methods.

TABLE 4.13
Properties of the Hybrid Nonisocyanate Polyurethane Floor Covering ECPU 2851

Properties	Indices
Coating material	Two-component hybrid on isocyanate polyurethane
Ratio of components on weight (binder:hardener)	100 : 55
Viscosity of binder	No more 3000 cps at 25°C (Brookfield, Spindle 29, velocity 100rpm)
Standard colors	According to catalog of basic colors
Pot life	40–60 min at 25°C
Substrate	Concrete, cement cover, asphalt, wood, etc.
Primer	Depends on substrate
Thickness of the coating	0.5–3 mm
Application temperature	15°C–25°C
Curing time at: 25°C	
• Light traffic	24 hours
• Full chemical load	7 days

Source: Reprinted from O. Figovsky and D. Beilin, "Building Materials Based on Advanced Polymer Matrix," *Journal Scientific Israel Advanced Technology* 10, no. 3 (2008): 1–119. With permission.

TABLE 4.14
Mechanical Properties of the Hybrid Nonisocyanate Polyurethane Floor Covering ECPU 2851 (after 10 days at 25°C)

Properties	Indices
Rupture strength	3–4 kg/mm^2
Elongation at rupture	4–8%
Hardness (Shore D)	80
Wear resistance	25 mg/1000 cycles
(Taber CS-17 wheel, 1000-g weight, ASTM 4060-90)	

Source: Reprinted from O. Figovsky and D. Beilin, "Building Materials Based on Advanced Polymer Matrix," *Journal Scientific Israel Advanced Technology* 10, no. 3 (2008): 1–119. With permission.

TABLE 4.15
Chemical Properties of the Hybrid Nonisocyanate Polyurethane Floor Covering ECPU 2851 (after 14 days at 25°C)

Media	Resistance
Water	
Petrol	
Aviation petrol	Resistant
Oil	
NaCl (5%)	
Formaldehyde	Resistant during short time
Sulfuric acid (20%)	
NaOH (20%)	
Benzene	Not Resistant
Alcohols	

Source: Reprinted from O. Figovsky and D. Beilin, "Building Materials Based on Advanced Polymer Matrix," *Journal Scientific Israel Advanced Technology* 10, no. 3 (2008): 1–119. With permission.

TABLE 4.16
Properties of Flooring and Coating Compositions Based on HNIPU

Substrate	Name	Main Property	Curing Time				Tensile Strength (kg/mm^2)	Ultimate Elongation (%)	Hardness[a]	Abrasion Resistance (TABER, mg/1000cycles)	Impact Resistance (kg*cm)	Flexibility (mm)	UV Resistance
			°C	Touch (h)	Light Traffic (h)	Full (day)							
1	2	3	4	5	6	7	8	9	10	11	12	13	14
							Floorings						
Concrete, Cement cover.	ECPU 2851 WC	Chemical -resistant indoor	25	3	24	7	3-4	4–8	80	25	20[b]		No
Asphalt, Wood	ECPU 2851 WCTC					7–10		4–10	70–75	25–30			
	ECPU 3570			6	48	10	3.9	9	70–80	25			No
Concrete, Cement cover. Wood	ECPU 2851 WCLS	Chemical-resistant indoor with increased light resistance	25	5	30	7–10	2.5–3	4–8	72	40–45			Yes
Concrete, Cement cover. Wood	ECPU 3770K	Cold cure 8°C–10°C indoor	8–10	18		14	2.5–3	10–20	70	25–30			No

1	2	3	4	5	6	7	8	9	10	11	12	13	14
							Coatings						
Metal, Ceramics, Plastic, Wood	ECPU 2994	Chemical-resistant indoor	25 110	4–6		7–10 0.1			H-2H		30	1–3	No
	ECPU 1510L		25 110	12–13		7–10 0.1			H		40	1–2	Yes
Ceramics, Plastic, Wood	ECPU 2896C		25 120	4–6		7–10 0.1	6–8	4–6	H-2H			1–3	No
Metal, Ceramics, Plastic, Wood	ECPU 2402 AC	Weather-resistant outdoor	110–120			0.1			H				Yes

Source: Reprinted from O. Figovsky and D. Beilin, "Building Materials Based on Advanced Polymer Matrix," *Journal Scientific Israel Advanced Technology* 10, no. 3 (2008): 1–119. With permission.

[a] For coatings and molding material: hardness according to Shore D; for paints: pencil hardness.

[b] Test data according to Polymer Institute, GmbH, Stuttgart, Germany for technical qualification of the product for the German flooring market.

REFERENCES

1. Thomson, T. *Polyurethanes as Specialty Chemicals: Principles and Applications*. Boca Raton, FL: CRC Press, 2005.
2. Meier-Westhues, U. *Polyurethanes: Coatings, Adhesives and Sealants*. Hanover, NH: Vincentz Network GmbH & Co KG, 2007.
3. Figovsky, O., et al. Soviet Union patent numbers: SU529197, 1976; SU563396, 1977; SU628125, 1978; SU630275, 1978; SU659588, 1979; SU671318, 1984; SU707258, 1984; SU903340, 1982; SU908769, 1982; SU1126569, 1984; SU 1754747, 1992; SU 1754748, 1992.
4. Figovsky, O., and Shapovalov L. "Cyclocarbonate-Based Polymers Including Non-Isocyanate Polyurethane Adhesives and Coatings." In *Encyclopedia of Surface and Colloid Science*, ed. P. Somasundaran, vol. 3, 1633–1653. New York: Taylor & Francis, 2006.
5. Leykin, A., Beilin, D., Birukova, O., Figovsky, O., and Shapovalov L. "Nonisocyanate Polyurethanes Based on Cyclic Carbonate: Chemistry and Application (review)," *Scientific Israel Technological Advantages* 11, nos. 3–4 (2009): 160–190.
6. Guan, J., Song, Y., Lin, Y., Yin, X., Zuo, M., Zhao, Y., Tao, X., and Zheng, Q. "Progress in Study of Non-isocyanate Polyurethane," *Ind. Eng. Chem. Res.* 50 (2011): 6517–6527.
7. Rappoport, L. Ya., Petrov, G. N., Trostyanskaya, I. I., and Gavrilova, O. P. "Polyurethane Elastomers Obtained without the Use of Diisocyanates," *Inter. Polymer Sci. Technol.* 8, no. 5 (1981): T/68–T/70.
8. Yagund, E. M., Maklakov, L. I., Stroganov, V. F., and Savchenko, V. N. "Studies of Hydrogen Bonds in Model Urethane Compounds Obtained by the 'Cyclocarbonate–Amine' Reaction," *Journal Appl. Spectroscopy* 45, no. 1 (1987): 737–741.
9. Tomita, H., Sanda, F., and Endo, T. "Structural Analysis of Polyhydroxyurethane Obtained by polyaddition of Bifunctional Five-Membered Cyclic Carbonate and Diamine Based on the Model Reaction," *Journal Polymer Sci. A* 39 (2001): 851 –859.
10. Zabalov, M. V., Tiger, R. P., and Berlin, A. A. "Reaction of Cyclocarbonates with Amines as an Alternative Route to Polyurethanes: A Quantum-Chemical Study of Reaction Mechanism," *Doklady Chemistry* 441, no. 2 (2011): 355–360.
11. Zabalov, M. V., Tiger, R. P., and Berlin, A. A. "Mechanism of Urethane Formation from Cyclocarbonates and Amines: A Quantum Chemical Study," *Russian Chemical Bulletin* 61, no. 3 (2012): 518–527.
12. Rokicki, G. Prog. Aliphatic Cyclic Carbonates and Spiroorthocarbonates bonates as Monomers Polym. Sci. 25, no. 259 (2000). PP 259–342.
13. North, M., Pasquale, R., and Young, C. "Synthesis of Cyclic Carbonates from Epoxides and CO_2," *Green Chem.* 12, no. 9 (2010): 1514–1539.
14. Pescarmona, P. P., and Taherimehr, M. "Challenges in the Catalytic Synthesis of Cyclic and Polymeric Carbonates from Epoxides and CO2," *Catal. Sci. Technol.* 2, no. 11 (2012): 2169–2187.
15. North, M. "Synthesis of Cyclic Carbonates from Epoxides and Carbon Dioxide Using Bimetallic Aluminium (Salen) Complexes," *ARKIVOC*, no. i (2012): 610–628.
16. Li, R., Tong, X., Li, X., and Hu, C. "Chlorine-Free Catalysts for Green Synthesis of Cyclic Carbonates from Carbon Dioxide," *Pure Appl. Chem.* 84, no. 3 (2012): 621–636.
17. Figovsky, O. "Hybrid Nonisocyanate Polyurethane Network Polymers and Composites Formed Therefrom", US Patent 6120905, Issued: September 19, 2000.
18. Figovsky, O., and Shapovalov, L. "Preparation of Oligomeric Cyclocarbonates and Their Use in Nonisocyanate Polyurethanes", US Patent 7232877, Issued: June 19, 2007.
19. Figovsky, O., Shapovalov, L., Blank, N., and Buslov, F. "Cyclocarbonate Groups Containing Hydroxylamine Oligomers from Epoxycyclocarbonates". US Patent 6407198, Issued: June 18, 2002.

20. Figovsky, O., Shapovalov, L., Blank, N., and Buslov, F. "Method of Synthesis of Polyfunctional Hydroxyurethane Oligomers and Hybrid Polymers Formed Therefrom". EP Patent 1070733, Issued: January 24, 2001.
21. Webster, D. C., and Crain, A. L. "Synthesis and Applications of Cyclic Carbonate Functional Polymers in Thermosetting Coatings," *Progress in Organic Coatings* (2000): 275–282.
22. Figovsky, O., "Improving the Protective Properties of Non-Metallic Corrosion-Resistant Materials and Coatings," *Journal of Mendeleev Chemical Society* 33, no. 3 (1998), 31–36.
23. Figovsky, O., and Shapovalov, L. "Nanostructured Hybrid Nonisocyanate Polyurethane Coatings," *Journal Paint and Coating Industry*, June 1, 2005.
24. Figovsky, O., and Shapovalov, L. "Hybrid Nonisocyanate Polyurethane Adhesives," *Proceedings of the International Conference Polymer Bonding 2004*, April 27–28, 2004, Munich, Germany, 99–103.
25. Birukov, O., Beilin, D., Figovsky, O., Leykin, A., and Shapovalov, L. Liquid Oligomer Composition Containing Hydroxy-Amine Adducts and Method of Manufacturing Thereof, US Patent 0144966 applied for in 2010.
26. Guillaume, S. M., and Carpentier, J-F. "Recent Advances in Metallo/Organo-Catalyzed Immortal Ring-Opening Polymerization of Cyclic Carbonates," *Catal. Sci. Technol.* 2 (2012): 898–906.
27. Figovsky, O., Birukov, O., Shapovalov, L., and Leykin, A. "Hydroxyurethane Modifier as Effective Additive for Epoxy Matrix," *Journal Scientific Israel Technological Advantages* 13, no. 4 (2011): 122–128.
28. Birukov, O., Figovsky, O., Leykin, A., Potashnikov, R., and Shapovalov, L. Method of Producing Hybrid Polyhydroxyurethane Network on a Base of Carbonated-Epoxidized Unsaturated Fatty Acid Triglycerides, US Patent 20120208967 applied for in 2012.
29. Couvert, D., Brosse, J. C., Chevalier, S., and Senet, J. P. "Monomeres acryliques a fonction carbonate cyclique, 2 Modification chimique de copolymeres a groupements carbonate cyclique lateraux," *Macromol. Chem.* 191 (1990): 1311–1319.
30. Moshinsky, L., and Figovsky, O. "Chemical Resistance of Epoxy Polymers Depending on Structure of Amino-Phenolic Hardeners," *Journal Scientific Israel Advanced Technology* 1, no. 1 (1999): 28–34.
31. Burgel, T., and Fedtke, M. "Reactions of Cyclic Carbonates with Amines: Model Studies for Curing Process," *Polym. Bull.* 27 (1991): 171–177.
32. Birukov, O., Beilin, D., Figovsky, O., Leykin, A., and Shapovalov, L. Nanostructured Hybrid Oligomer Composition. US Patent 7,820,779 B2, Issued: October 26, 2010.
33. Figovsky, O., Shapovalov, L., Leykin, A., Birukova, O., and Potashikova, R. "Recent Advances in the Development of Non-Isocyanate Polyurethanes Based on Cyclic Carbonates", *PU Magazine International*, no. 4 (2013) 256–265.
34. "A Novel Cyclocarbonate-Based Technology," *Polyurethane Newsletter* no. 68 (2004).
35. Garland Floor, http://www.garlandfloor.com/polyinfo.htm.
36. Figovsky, O. "Improving the Protective Properties of Nonmetallic Corrosion-Resistant Materials and Coatings," *Journal of Mendeleyev Chemical Society*, no. 3 (1988): 31–36.
37. Figovsky, O., and Shapovalov, L. "Monolithic Flooring Based on Nonisocyanate Polyurethane Binders," *Proceedings of the 6th International Colloquium, Industrial Floors '07*, vol. 2, Germany, January, 2007, 693–698.
38. Figovsky, O., and Beilin, D. "Building Materials Based on Advanced Polymer Matrix," *Journal Scientific Israel Advanced Technology* 10, no. 3 (2008): 1–119.
39. Zajc, A., Courad, L., Garbacz, A., and Wolf, L. "Testing Procedures and Other Regulations," RILEM TC 184-IEF *Industrial Floors: State-the-Art Report*, 2006, 101–110.
40. Stinner, N., and Ukrainczyk, V. "'Zagreb' Test According to N. Stinner and V. Ukrainczyk: Test Impact Resistance of Industrial Floors Overlays," *Proceedings of Industrial Floors '03, 5th International Colloquium*, 2003, 627–637.

41. Linsbauer, H. N., and Tschegg, E. K. "WST Method According to H. N. Linsbauer and E. K. Tschegg: Die Bestimmung der Bruchenergie von Zementgebundenen Werkstoffe an Würfelproben," *Zement und Beton* 31, 1986.
42. Delebecq, E., Pascault, J-P., Boutevin, B., and Ganachaud, F. "On the Versatility of Urethane/Urea Bonds: Reversibility, Blocked Isocyanate, and Non-Isocyanate Polyurethane," *Chem. Rev*. 113 (2013): 80–118.
43. Figovsky, O., Shapovalov, L., Leykin, A., Birukova, O., and Potesshnikova, R., "Nonisocyanate Polyurethanes Based on Cyclic Carbonates and Nanostructures Composites," *Journal Scientific Israel Technology Advanced* 14, no. 4 (2012): 95–106.
44. Phazer Inc., http://www.phazerinc.com/phazerthane.html.
45. Nanotech Industries, http://nanotechindustriesinc.com/GPU.php.

ASTM STANDARDS CITED IN CHAPTER 4

1. **C 518** Standard Test Method for Steady-State Thermal Transmission Properties by Means of the Heat Flow Meter Apparatus
2. **D 523** Standard Test Method for Specular Gloss
3. **D 1622** Standard Test Method for Apparent Density of Rigid Cellular Plastics
4. **D 2196** Standard Test Methods for Rheological Properties of Non-Newtonian Materials by Rotational (Brookfield type) Viscometer
5. **D 2369** Standard Test Method for Volatile Content of Coatings
6. **D 3359-07,B** Standard Test Methods for Measuring Adhesion by Tape Test
7. **D 3363-05** Standard Test Method for Film Hardness by Pencil Test
8. **D 4060** Standard Test Method for Abrasion Resistance of Organic Coatings by the Taber Abraser
9. **D 5402-06** Standard Practice for Assessing the Solvent Resistance of Organic Coatings Using Solvent Rubs

5 Crack-Resistant and Anticorrosive Coatings Based on Vulcanized Water Dispersion of Chlorosulfonated Polyethylene

INTRODUCTION

Protective crack-resistant coatings, based on chlorosulfonated polyethylene (CSPE), are widely known due to their unique crack- and corrosion-resistant properties as compared to other rubbers. These coatings also have high ozone, weather, and wear resistance, as well as excellent mechanical and dielectric characteristics.

CSPE-based compositions for coatings are traditionally produced with application of aromatic solvents (xylene, toluene, etc.) characterized by high toxicity and explosion and fire hazards. These properties interfere with application of conventional coatings. Moreover, the polymer concentration in such compositions cannot exceed 20%.

Metal oxides (mostly lead oxides), primary aliphatic and aromatic polyamines, epoxy resins, low-degree polyamide resins, amino–epoxy adducts, and polyfunctional isocyanates are commonly used as CSPE vulcanizing agents. However, such vulcanizates have many disadvantages, for example, metal oxides are not soluble in solvents, and primary aliphatic polyamines are characterized by high toxicity and volatility. Furthermore, the use of aliphatic polyamines initiates spontaneous mixture vulcanization, and the application of aromatic polyamines and aromatic-based amino–epoxy adducts gives rise to the black color of the vulcanizates. Epoxy resins have a low coat-forming ability, and they are not water soluble. This disadvantage is also typical for low-degree polyamide resins.

For the first time, we have succeeded in preparing a vulcanized waterborne protective crack-resistant coating—based on CSPE for concrete and other substrates—that has better corrosion, crack, gas, and water resistance. These properties have been achieved by the application of a waterborne vulcanizate, which has never been used before. For this purpose, compounds such as Mannich alkalis (MAs) were applied, which cause the formation of quaternary ammonium bonds in a vulcanized net of the saturated elastomer and have reactive functional groups.

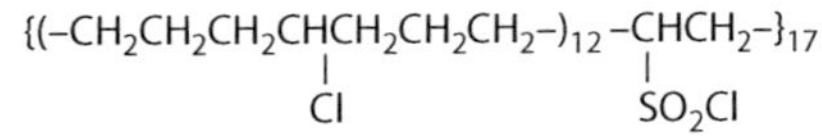

STRUCTURE 5.1 Chlorosulfonated polyethylene. (Reprinted from O. Figovsky, V. Karchevky, and D. Beilin, "Protective Crack-Resistant Waterborne Coatings Based on Vulcanized Chlorine-Sulpho-Polyethylene," *Scientific Israel Technological Advantages* 3, nos. 1–2 (2001). With permission.)

It is known that MAs obtained by the result of a Mannich reaction of phenols, formaldehydes, and acryls are good water and organic solvent soluble materials, and are nontoxic and nonvolatile. The application of MAs as a structure-forming component of CSPE makes it possible to manufacture ecologically safe vulcanized coatings for any substrate. These coatings can be used in aircraft, automotive, shipbuilding, paint and varnish industries, civil engineering, and so on as crack- and corrosion-resistant materials.

The new material composition on the CSPE base with Structure 5.1 and vulcanized by a water solution of MAs provides the formation of a saturated polymer vulcanized net. This makes it possible to obtain impenetrable crack-resistant coatings for metal, concrete, plastic, and other substrates, protecting them from damaging effects of the environment.

As a result, CSPE-based vulcanized waterborne crack-resistant coatings are effective if used as follows:

- Underground and ground-level hydro-isolation of buildings for corrosive protection under aggressive media attack
- Crack-resistant roof covers for various climate zones
- Outer-surface coatings in a wide range of colors characterized by high UV-resistance for buildings and other structures
- Protective coatings for coastal and underwater concrete structures and installations

COATING COMPOSITION [1–5]

Test samples of coatings, 0.3–0.4 mm thick, were prepared on the CSPE water dispersion base hardened by MAs at room temperature, with the addition of a filler, pigments, and modifiers, for steel and concrete substrates. To achieve the optimal composition model, a number of mechanical and physical-chemical experiments were executed. The tests were performed in the nodal points of a two-level full factor experiment. The efficiency functions of the coating tests on metal substrates were adhesion strength and corrosion stability in acid and alkali media. The efficiency function of the tests on the concrete samples was deformation at tension, which is a controlling factor for the crack-resistance of a coating, on the basis that the tensile strength of the coating should be greater than is its adhesion strength to concrete.

We have investigated the two kinds of CSPE water dispersions: high and low density, and four types of vulcanizing agents (MAs):

1. 2,4,6 –tris(dimethylaminomethyl) phenol

$$\mathbf{HO{-}C_6H_2[CH_2N(CH_3)_2]_3}$$

2. 2-(dimethylaminomethyl) phenol

$$HO–C_6H_4–CH_2N(CH_3)_2$$

3. 2,6-bis(dimethylaminomethyl) cresol

$$HO–CH_3–C_6H_2[CH_2N(CH_3)_2]_2$$

4. 2,6-bis(dimethylaminomethyl)-4 tretbuthylphenol

$$HO–C(CH_3)_3–C_6H_2[CH_2N(CH_3)_2]_2$$

It was found experimentally that the optimal vulcanizing agent for crack-resistant and anticorrosive coating compositions was the *first* agent listed, which contained three amine groups of Structure 5.2: two are arranged in the ortho position, and one in the para position relative to phenol hydroxide. Such a structure of the vulcanizing agent provides the maximal cross-linking of elastomer, that is, its vulcanization, and makes it possible to obtain coatings that are characterized by high strength, elasticity, and resistance to attack by corrosive media. It is noteworthy that the other types of MA have one or two amine groups only in the ortho position.

Five compositions of the coating were obtained:

- *Crack-resistant coatings* with the high elongation for concrete and other substrates that make their favorable to brittle destruction
- *Anticorrosive coatings* with high adhesion to a substratum and high rupture strength
- *Architectural coatings for walls* with low water absorption and high resistance to UV exposure
- *Architectural coatings for floors* with high wear resistance, keeping a natural texture of a substratum
- *Temporary (detachable) coatings* with extremely low adhesion to a substrate

The optimal compositions for these coatings are illustrated in Table 5.1.

The composition was prepared as follows:

- Water dispersions of CSPE emulsion are mixed thoroughly with all composition components excluding Mannich alkalis.
- The second main component (i.e., water solution of Mannich alkalis) was added to the mixture just before applying the coating to the surface of the substrate.

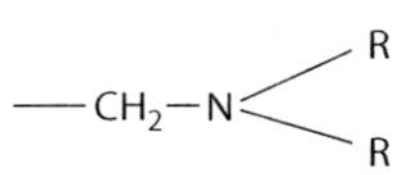

STRUCTURE 5.2 Tertiary amine. (Reprinted from O. Figovsky, V. Karchevky, and D. Beilin, "Protective Crack-Resistant Waterborne Coatings Based on Vulcanized Chlorine-Sulpho-Polyethylene," *Scientific Israel Technological Advantages* 3, no. 1–2 (2001). With permission.)

TABLE 5.1
Optimal Composition of Waterborne CSPE Coatings

	Specific Character				
Component (%)	**Crack-Resistant**	**Anticorrosive**	**Architectural for Walls**	**Architectural for Floors**	**Temporary**
CSPE Emulsion (HYP-605®)	19.6	22	20	22	22
Mannich alkalis	1.6	2.5	0.5	1.6	1.2
Baypren MKB®	30.5	—	—	—	25
n-Butanol	2.2	—	—	—	1.3
Chlorinated-paraffin emulsion	—	—	1	—	—
Polyvinyl alcohol	—	—	—	—	1.5
Setalux 8552 AQ-50®	10.8	26.8	12.5	22	—
Setalux 8552 AQ-36®	5.4	—	16.3	22	—
Aerosil®	—	—	—	1	1
Graphite	—	—	—	0.4	—
Talc	—	—	—	3	19
Magnesium oxide	2.2	0.8	1	—	—
Zinc oxide	2.2	0.8	1	—	—
Barium sulfate	—	5	2.5	—	—
Titanium oxide	—	—	10	—	—
Chalk	—	—	20	—	—
Pigment inorganic	2.7	7.7	—	—	—
Filler (andesite/stone/diabas powder)	10.8	19	—	7	—
Foam inhibitor (DEEFO PI-40®)	—	—	0.2	—	—
Water	12	15.4	15	21	29

- After all components were mixed for 5–10 minutes, the composition was ready for use as a coating.

The coatings could be applied on a substrate by any suitable method, for example, by brushing, roller, spraying, and so on. Several layers of the coating can be applied, and each layer should be dried for 6–8 hours at ambient temperature (though not lower than 15°C).

PHYSICAL–MECHANICAL PROPERTIES AND CORROSION RESISTANCE OF THE VULCANIZED CSPE COATING [6–8]

Physical–Mechanical Characteristics

The most important physical–mechanical characteristics of the coatings are shown in Table 5.2. Typical diagrams of the stress–strain relationship for coatings with optimal composition on concrete substratum are plotted in Figures 5.1 and 5.2.

TABLE 5.2
Physical–Mechanical Properties of Waterborne CSPE Coatings

		Waterborne CSPE Vulcanized by Water Solution of Manniach Acid				
Properties	**ASTM Standard**	**Crack-Resistant**	**Anticorrosive**	**For Walls**	**For Floors**	**Temporary**
1	2	3	4	5	6	7
Curing		24 hours @ 25°C Or 0.5 hours @ 90°C				
Coverage ft²/gallon per 1 mil of dry film thickness		750	650–700	700	650	800–900
Solid content						
Weight,%		52–55	45–50	58–60	40–45	44–45
Volume,%		43–45	30–35	45–50	30–32	37–40
Crack resistance,[a] mm		3	—	—	—	—
UV exposure (Cycle A, ASTM D-4329)	D4329	No more than 3% No visible changes				
Change of mechanical properties	G151 G154					
Color fastness						
Viscosity, cps @ 25°C, shp. 21, 100 rpm	D2196	650–700	500–600	300–400	450–500	150–300
Dielectric constant, ohm*cm	D257	—	10^9–10^{10}	10^9–10^{10}	—	—
Water absorption,%	D471	4–6	3–4	4–5	2–3	6–8
Flash point,°C	D56	Noncombustible				
Standard color		Depend on pigments				
Toxicity		Nontoxic				
V.O.C.		Noncombustible, nontoxic, nonexplosive				
Adhesion, MPa	D1002	6.2 to metal, concrete	20 to metal	6 to concrete	5–6 to metal, concrete	0–0.3 to aluminum, steel
Ultimate strength at tension, MPa	D412	8	18	4	10–11	2–3
Elongation,%		320	80	25	8	20–30
Abrasive strength, wear cycles/mil	D-4060	$3*10^3$	$1.5*10^3$	$0.5*10^3$	$1*10^3$	—
Hardness, Shore D	D-2240	67–70	78	83	80–83	40
Impact	D-2794	70	50	50	70	30

[a] Crack-resistance criterion of elastomeric coating is maximal width of an overlapped crack in concrete substratum at which coating ruptures.

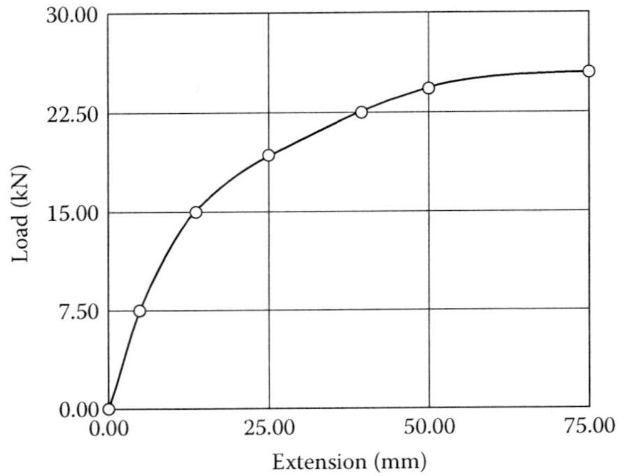

FIGURE 5.1 Typical tensile load–extension curve for a coating with optimal composition on concrete substrate (ASTM D412). (Reprinted from O. Figovsky, V. Karchevky, and D. Beilin, "Protective Crack-Resistant Waterborne Coatings Based on Vulcanized Chlorine-Sulpho-Polyethylene," *Scientific Israel Technological Advantages* 3, nos. 1–2 (2001). With permission.)

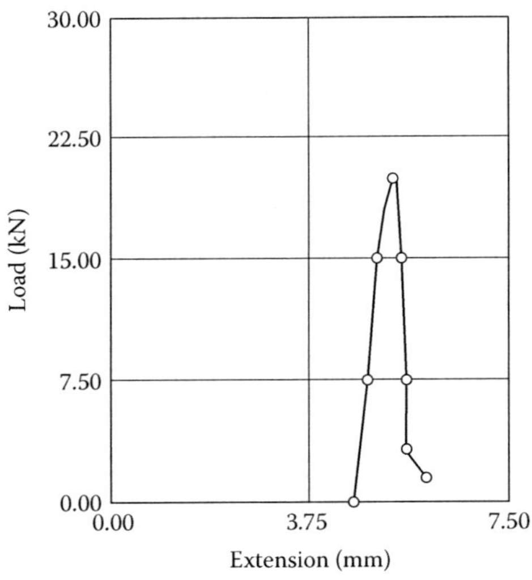

FIGURE 5.2 The tensile load–shear strain relationship of the coating with optimal composition on concrete substrate (ASTM 1002). (Reprinted from O. Figovsky, V. Karchevky, and D. Beilin, "Protective Crack-Resistant Waterborne Coatings Based on Vulcanized Chlorine-Sulpho-Polyethylene," *Scientific Israel Technological Advantages* 3, no. 1–2 (2001). With permission.)

Resistance to UV Radiation

A series of accelerated exposure tests was performed to investigate the effect of UV radiation on the coating material, the stress–strain state in tension and shear, and the color fastness of the coating material. The tests were carried out according to ASTM Standards ASTM D4329, G151 and G154.

The UV light source was the UVA 340 lamp with spectral characteristics in a range of wavelengths corresponding to a spectrum of sunlight, and an emission peak of 343 nm. Typical lamp irradiance is 0.77 $W/m^2/nm$.

The lamp was set in the test chamber, which was equipped with a temperature and radiometric control system. The specified uninsulated black panel sensor was used for measurement of temperature. The sensor was mounted on a support within the specimen exposure area so that it received the same radiation and cooling conditions as the test samples. The ambient temperature at a distance of 150 mm from the chamber was maintained within the 18°C–23°C range. Ventilation and air conditioning systems eliminated superfluous heat and humidity.

Samples of two types were made: Type KB (crack-resistant) for concrete and ceramic substrates, and type KM (anticorrosive) for metal and other substrates. Three series of various color samples (white, yellow, and green) were submitted on three substrate samples in each test series. All samples had a rectangular form, and measured 120 × 120 × 0.5 mm in size. Samples were fixed on a contour in special stainless steel frames, and used an aluminum substrate with a thickness of 0.635 mm. Samples were exposed in a vertical orientation.

The UV test exposure procedure corresponded to the cycle "A" [4] and provided an 8-hour UV exposure test with uninsulated black panel temperature control at 60±3°C.

Test conditions and results (obtained on average values of tested sample characteristics) are illustrated in Table 5.3.

TABLE 5.3
Mechanical Properties of Waterborne CSPE Coatings with and without UV Exposure

		Coating Type			
		Anticorrosive (KM)		Crack-Resistant (KB)	
Characteristics	ASTM **Standard**	**Without UV Exposure**	**After UV Exposure**	**Without UV Exposure**	**After UV Exposure**
Lamp irradiance ($W/m^2/nm.$)	G154	—	0.77	—	0.77
Temperature (°C)	D5031	+21	+58	+22	+61
Shear strength (MPa)	D1002	20.7	19.8	6.0	5.9
Ultimate strength at tension (MPa)	D412	18.1	19.2	8.1	9.4
Elongation (%)		68	66	302	290
Color:	D1729	White Green Yellow	No visible change	White Green Yellow	No visible change

The fluorescent UV exposure accelerated tests of anticorrosive and crack-resistant coating samples showed that the following:

1. Ultimate strength at tension of samples after UV exposure was above the control samples on an average of about 3%, due to the influence of light energy on the cross-link process of the polymeric composition of the coating.
2. The influence of UV exposure on the adhesive strength of the coatings is practically insignificant.
3. Change in the color of the coating sample under the influence of UV exposure is visually imperceptible.

Dielectric Characteristics

Research on the dielectric characteristics of coating films was performed using a teraohmmeter. Five samples of different coating compositions were examined on a metal substrate. Samples were prepared on the basis of two types of water dispersion rubbers, CSM-200 and CSM-400. The test samples were placed in a drying oven. The electrical resistance of the samples was measured via thermocouple wires that were connected to the teraohmmeter. The results of tests are shown on Figures 5.3 and 5.4.

Analysis of the results allowed conclusions to be drawn from the high-resistance measurements of the coatings. The mean volumetric electrical resistance of the samples tested was $\rho_v = 10^9$–10^{10} ohm.cm.

Corrosion Resistance

Corrosion resistance tests were carried out in accordance with the ASTM D1654 requirements using the following aggressive media: 30% sulfuric acid, 20% hydrochloric acid, or 30% potassium hydroxide. Coating samples were kept in this media for 250 hours at 40°C, and were then subjected to tensile load tests. The results of the tests are shown in Table 5.4.

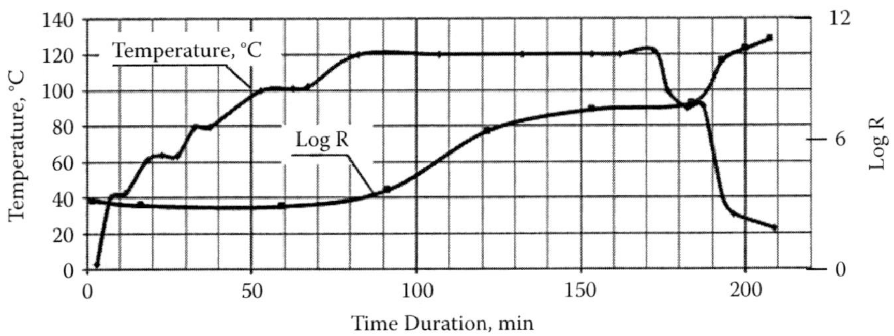

FIGURE 5.3 Changes in the electrical resistance of a coating sample on the CSM-200 base during hardening.

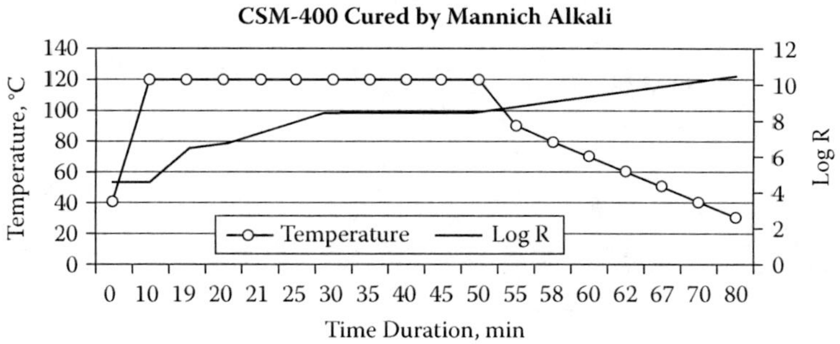

FIGURE 5.4 Changes in the electrical resistance of a coating sample on the CSM-400 base during hardening.

TABLE 5.4
Results of Corrosion Tests

Component	ASTM Standard	Coating Type: Anticorrosive	Coating Type: Crack-Resistant
Resistance in 30% Sulphuric Acid			
Change of coating weight after testing (%)	D-1654	+3.2	+6.4
Coefficient corrosion resistance for strength[a]	D-412	0.93	0.82
Coefficient corrosion resistance for elongation[b]		0.95	0.84
Resistance in 20% Hydrochloric Acid			
Change of coating weight after testing (%)	D-1654	+4.1	+6.9
Coefficient corrosion resistance for strength[a]	D-412	0.91	0.89
Coefficient corrosion resistance for elongation[b]		0.97	0.86
Resistance in 30% KOH			
Change of coating weight after testing (%)	D-1654	+2.4	+4.8
Coefficient corrosion resistance for strength[a]	D-412	0.95	0.93
Coefficient corrosion resistance for elongation[b]		0.97	0.96

[a] Rupture load of the coating sample after test in corrosive medium/Rupture load before corrosion test.

[b] Elongation of the coating sample at rupture after test in corrosive medium/Elongation of the coating sample at rupture before corrosive test.

TABLE 5.5
Results of Long-Term Degradation Tests

Medium	NaOH 30%	NaCl 10%	Ethyl Alcohol	Machine Oil	HNO_3 20%	HNO_3 10%	Acetic Acid 100%
Change of coating weight after testing (%)	+3.94	−1.99	−2.6	+3.6	+4.6	+3.5	−2.2

In addition, long-term degradation experiments were performed on free films. The film samples were placed in corrosive media for 3 months. The results are illustrated in Table 5.5.

APPLICATION OF CSPE COATING FOR FORMING CONCRETE AND REINFORCED CONCRETE STRUCTURES [9–11]

Film coatings are widely applied in forming concrete and reinforced concrete structures to accelerate concrete hardening and reduce crack appearance at setting. The film prevents evaporation of the water contained in the concrete, promotes hydration of the cement, and thus provides reliable contact of the cement stone and filler. It promotes an increase in concrete strength up of up to 30%, reduces damage and time of processing, and restricts moisture absorption.

The developed composition of a high-adhesion CSPE coating cured by a water solution of MA was applied as a covering in the formation of concrete articles. It was investigated with respect to its setting and hardening processes by evaluating the concrete under loading.

According to the theory of V. Solomatov and V. Seljaev [8], shrinkage stresses in a concrete coating create the compressive stresses σ_{con} in contact layer of a concrete substratum. Therefore, the value of the moment M_{crack} carried by a section before crack formation in a substratum is increased and it is possible to determine a formula for prestressed reinforced concrete.

$$M_{crack} = R_{con} W_{crack} + M_{cor} \tag{5.1}$$

where R_{con} is the ultimate tensile strength of concrete.

Obviously, the forces arising from shrinkage strains are transmitted on the concrete in a plane of contact coating–substratum for an element with a rectangular cross section (Figure 5.5).

$$M_{cor} = N_{shr}\,(e_o + r_{core});\; e_o = 0.5h_{con};\; r_{cor} = h_{con}/6;\; M_{cor} = 2/3(N_{shr}\,h_{con})$$

Taking into account that $N_{shr} = \sigma_{shr}\,\delta_c b$ and $W_{crack} = (0.292 + n\mu)bh^2_{con}$ where σ_{shr} is the shrinkage stress averaged on the height of the polymer coating, and μ is the coefficient of reinforcement we have:

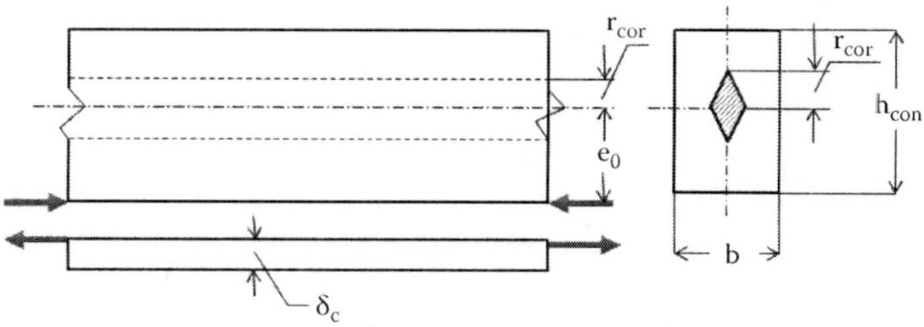

FIGURE 5.5 Calculation of polymer coating shrinkage strains. (Reprinted from O. Figovsky, V. Karchevsky, and D. Beilin, "Application of Polymeric Coatings for Concrete Structures," *Scientific Israel Technological Advantages* 5, nos. 1–2 (2003): 84–92. With permission.)

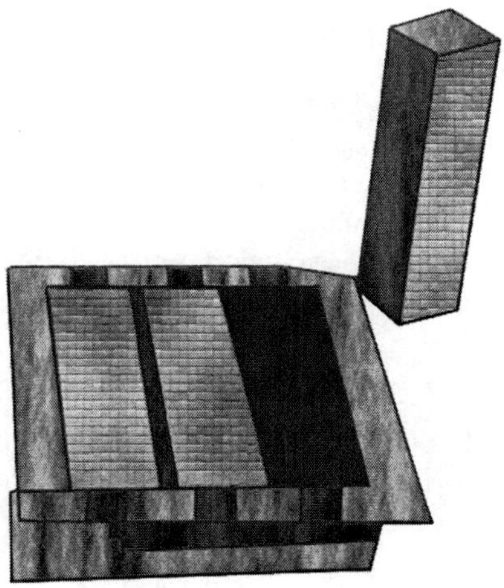

FIGURE 5.6 Concrete samples with a coating during natural hardening. (Reprinted from O. Figovsky, V. Karchevsky, and D. Beilin, "Application of Polymeric Coatings for Concrete Structures," *Scientific Israel Technological Advantages* 5, nos. 1–2 (2003): 84–92. With permission.)

$$M_{crack} = R_{con} W_{crack} \left[1 + 2\sigma_{shr} \delta_c / 3R_{con} h_{con} (0.292 + n\mu)\right], \tag{5.2}$$

where $n = 2E_{rf}/E_{con}$, E_{rf} is the modulus of elasticity of reinforcement, and E_{con} is the modulus of elasticity of concrete.

From the Equation (5.2) it follows that with an increase of shrinkage stresses and thickness of a polymeric coating, the moment of crack formation for reinforced concrete beams increases.

To check the analytical dependence conformity to experimental results, a series of reinforced concrete beams measuring 160 × 40 × 40 mm, reinforced by one bar Ø 6 mm was prepared. After vibrating, the coating is applied to a surface of a ready concrete mixture, following which curing takes place in natural conditions (Figure 5.6).

By the third day after molding, a beam had a curve *f* due to coating shrinkage. On value of this curve, the shrinkage stress σ_{shr} that deformed the beam was determined:

TABLE 5.6
Experimental Results

Beam Number	δ_c/h_{con}	f (μm)	σ_{sh} (MPa)	M_{crack} (%)
Beam 0 (without coating)	0	0	0	100
Beam 1	0.125	2	6.7	112
Beam 2	0.150	3	8.3	115
Beam 3	0.175	3.5	8.3	117
Beam 4	0.200	4	8.3	119

$$\sigma_{shr} = 4\,E_{con}\,h^2_{con}\,f/3\delta_c L^2 \tag{5.3}$$

where L is the span of the beam sample. Further, by Equation (5.2) the crack formation moment was found.

The samples were tested for a bend by concentrated force in the middle span of the beam with a polymeric coating in a tension zone. During the experiment, the bending moment appropriate to appearance of the first visible crack was registered. From the analysis of experimental results illustrated in Table 5.6, it follows that due to the shrinkage of the polymeric coating, the bending moment crack formation of the concrete substratum is increased by 12%–19%.

PHENOMENOLOGICAL MODEL OF CRACK-RESISTANT COATINGS FOR CONCRETE SUBSTRATES [2,8,12–18]

The application of polymeric film coatings to reinforced concrete structures not only allows reliable isolation of the concrete and reinforcement against aggressive attacks, but also provides a positive effect on the stress–strain properties of the structure. As has been claimed in [2,13,14] that elastomerically hardened coatings can increase the crack resistance, load-bearing capability, and durability of reinforced concrete structures.

The crack resistance of a coating is defined by the following polymer characteristics: ultimate extensibility, shrinkage, tensile strength, modulus of elasticity, adhesion, the structure of the surface of the concrete substrate, and the stressed state of the coating. The crack-resistance criterion of elastomeric coatings is determined as the maximum width of an overlapped crack in the concrete at which the coating will rupture. Thus, the ultimate extension of the coating material at a tensile load, which is more than its adhesion to concrete, is the main factor that determines the crack resistance of a coating.

The stress state of a coating is closely interconnected with the adhesion of that elastomer to the concrete. If adhesion forces are not zero, "peeling" stresses arise in the coating, which is in a *plain stress* state. Such stresses can reach a critical value and concentrate in a zone of a concrete substrate where the cracks are formed. In the absence of adhesion, the coating functions independently of the substrate, and any increase in the exterior load will result in the development of a peeling area. Taking into consideration that the ultimate tension stress of any coating exceeds the adhesion stress, a peeled elastomeric film will stretch and overlap the cracks formed in the concrete substrate (Figure 5.7).

In order to estimate the efficiency of a protective crack-resistant coating during the formation and development of cracks in a concrete substrate, an experimental investigation and numerical analysis of the stress–strain state of a concrete sample with such coating at a tension load was carried out.

Three stages of the model under loading were considered:

- Before formation of a crack
- Formation of the first crack
- With the given crack width

The design model is generated in the frameworks of the finite element method. It represents a set of the concrete elements connected among themselves by elastic connections (Figure 5.8).

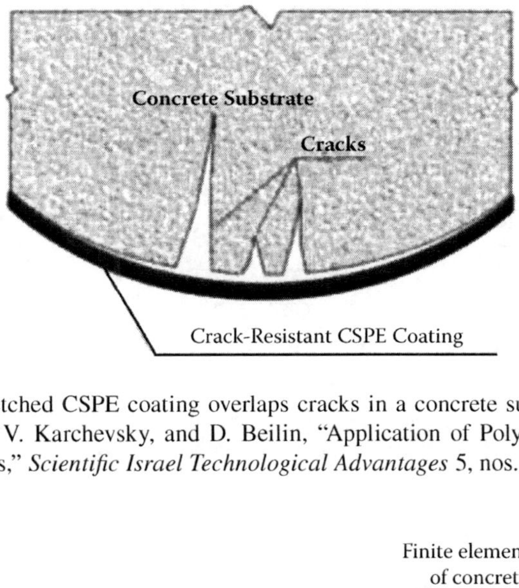

FIGURE 5.7 Stretched CSPE coating overlaps cracks in a concrete substrate. (Reprinted from O. Figovsky, V. Karchevsky, and D. Beilin, "Application of Polymeric Coatings for Concrete Structures," *Scientific Israel Technological Advantages* 5, nos. 1–2 (2003): 84–92. With permission.)

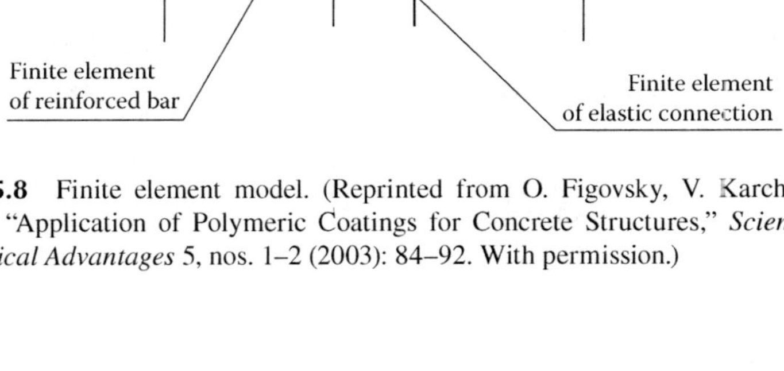

FIGURE 5.8 Finite element model. (Reprinted from O. Figovsky, V. Karchevsky, and D. Beilin, "Application of Polymeric Coatings for Concrete Structures," *Scientific Israel Technological Advantages* 5, nos. 1–2 (2003): 84–92. With permission.)

A central reinforced bar is modeled by the elastic one-dimensional finite element taking into account only deformations of compression and tension. External force is applied to the reinforced element.

At the *first* stage we determined the tension load corresponding to ultimate tension stress 2.7 (for concrete of class B-30). In so doing, the external tension load was 0.355 t, and crack width is 0.056 mm.

At the *second* loading stage, the connection that had reached a limiting tension stress and exhausted its capacity was excluded from the model. New calculations, at which there was a redistribution of forces in the remaining concrete element connections, were carried out. At this stage, displacement of the medial element located on an external surface of a sample was 0.0603 mm.

At the *third* stage, the stress–strain state of a sample at a crack width appropriate to a design tension stress of reinforced bar (340 MPa) was checked. This stress corresponds to external tension force of 1.710 t. Crack width was 1.77 mm. A diagram of the finite elements of the sample–model displacements is shown in Figure 5.9.

In this way, all stages of loading of the design model can be shown in a diagram (Figure 5.10).

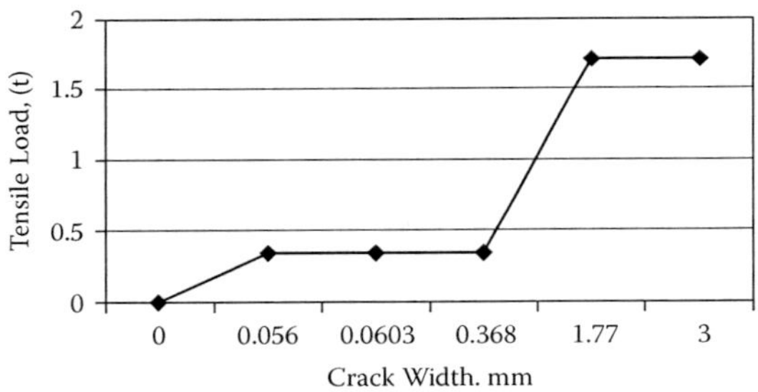

FIGURE 5.9 Displacement of finite elements of model at design tension stress in reinforced bar. (Reprinted from O. Figovsky, V. Karchevsky, and D. Beilin, "Application of Polymeric Coatings for Concrete Structures," *Scientific Israel Technological Advantages* 5, nos. 1–2 (2003): 84–92. With permission.)

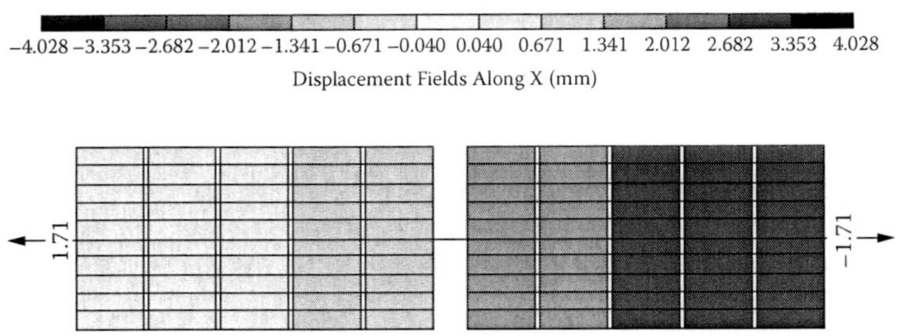

FIGURE 5.10 Development of crack formation in concrete substratum. (Reprinted from O. Figovsky, V. Karchevsky, and D. Beilin, "Application of Polymeric Coatings for Concrete Structures," *Scientific Israel Technological Advantages* 5, nos. 1–2 (2003): 84–92. With permission.)

Prismatic test samples measuring 160 × 40 × 40 mm in size were fabricated from concrete of class B-30, which was reinforced by one 8-mm diameter steel bar along the prism axis (Figure 5.11). Samples were tested in tension by means of a LLOYD LR 50K press. The process of crack formation and development was detected with the help of a microscope.

The crack resistance of the optimum coating composition of a concrete substrate was 2.0–2.5 mm in extension. Elongation of the coating before the moment of its fracture was determined as

$$\varepsilon_c = [1 - (\Delta_c - \delta_{cr}/\Delta_c)]100\% \tag{5.4}$$

where Δ_c, δ_{cr} were the length of the peeling zone of the coating and the maximum width of the crack in concrete substrate before the coating ruptured, respectively. It was found experimentally that the elongation $\varepsilon_c \approx 36\%$. The coating crack resistance factor, which is the relation of the crack resistance to the coating thickness, was ~ 2.4.

These test results are illustrated in Figure 5.12 and Table 5.7.

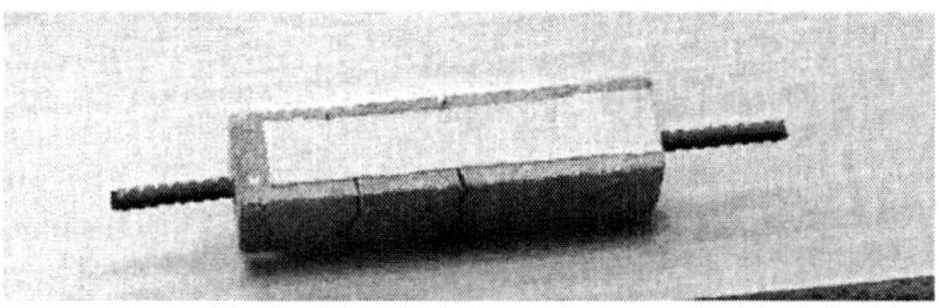

FIGURE 5.11 Test sample of crack-resistant coating on concrete substratum. (Reprinted from O. Figovsky, V. Karchevsky, and D. Beilin, "Application of Polymeric Coatings for Concrete Structures," *Scientific Israel Technological Advantages* 5, nos. 1–2 (2003): 84–92. With permission.)

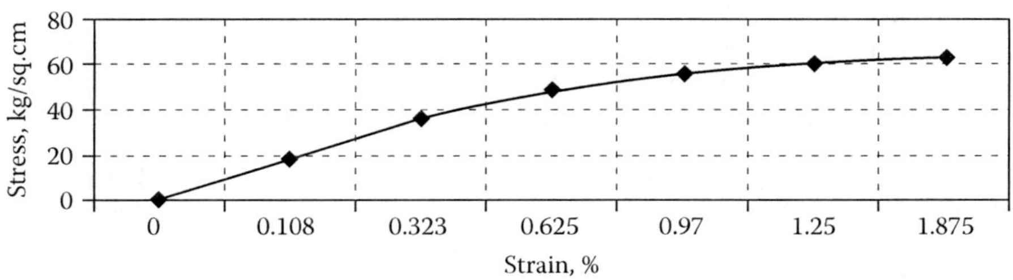

FIGURE 5.12 Stress–strain relationship of crack-resistant coating under tensile loading. (Reprinted from O. Figovsky, V. Karchevsky, and D. Beilin, "Application of Polymeric Coatings for Concrete Structures," *Scientific Israel Technological Advantages* 5, nos. 1–2 (2003): 84–92. With permission.)

TABLE 5.7
Stress–Strain State of Concrete Sample with Crack-Resistant Coating at Tension

Force (N)	0	7.5	15	19.285	22.5	24.2	25.285	32.8
Displacement (mm)	0	4.31	12.93	25	38.8	50	75	120.8
Stress (кg/cm^2)	0	18.5	37.04	47.65	55.56	59.75	62.47	81
Strain (%)	0	10.8	32.3	62.5	97	125	187.5	302

Source: Reprinted from O. Figovsky, V. Karchevsky, and D. Beilin, "Application of Polymeric Coatings for Concrete Structures," *Scientific Israel Technological Advantages* 5, nos. 1–2 (2003): 84–92. With permission.

SUMMARY

- A new type of anticorrosive and crack-resistant coating that is based on an ecologically safe CSPE dispersion hardened by aqueous solution of Mannich alkalis was developed and investigated.
- These new coating materials are nontoxic and should have a wide range of application.

REFERENCES

1. Karchevsky, V., and Figovsky, O. Coating Composition. US Patent Application 20-03/0078336A1US.
2. Figovsky, O., Karchevsky, V., and Beilin, D. "Crack-Resistant and Anticorrosive Coatings Based on Vulcanized Water Dispersion of Chlorine-Sulpho-Polyethylene," *Journal Anti-Corrosion Methods and Materials* 50, no. 2 (2003).
3. Figovsky, O., Karchevsky, V., and Beilin, D, "Protective Crack-Resistant Waterborne Coatings Based on Vulcanized Chlorine-Sulpho-Polyethylene," *Scientific Israel Technological Advantages* 3, no. 1–2 (2001).
4. Shneiderova. V. *Anticorrosive Varnish-Painted Coating in Construction*, Moscow: Stroiizdat, 1980 (in Russian).
5. Figovsky, O., Karchevsky, V., and Beilin, D., "Advanced Crack-Resistant Coatings Based on Water Dispersion CSPE," *The 5th Biannual Conference of the Corrosion Forum NACE Israel, Abstracts*, Tel-Aviv, Israel, 2002, 2.
6. Beilin, D., and Figovsky, O. "Stress–Strain State of Polymer Chemical-Resistant Coatings Curing and under Operating Conditions," *J. Composite Structures* 31 (1995): 253–256.
7. Figovsky, O., Karchevsky, V., and Beilin, D. "Stress–Strain State of Elastomeric Crack-Resistant Coatings for Concrete Substrate." In *Book of Abstracts of Twelfth International Conference Mechanics of Composite Materials,* Riga, Latvia, 2002, 29.
8. Solomatov, V., and Seljaev, V. *Chemical Strength of Composite Building Materials*, Moscow: Stroiizdat, 1987 (in Russian).
9. Weiss, W., Yang, W., and Shah. S. "Shrinkage Cracking of Restrained Concrete Slabs," *J. of Engineer Mechanics Div.*. ASCE 124, no. 7 (1998): 765–774.
10. Figovsky, O., Karchevsky, V., and Beilin, D. "Investigation of Shrinkage Stresses Developed at Setting of the Concrete Bearing Polymer Film Coatings," *Proceedings of the Third International Research Seminar Self-Desiccation and Its Importance in Concrete Technology*, Lund, Sweden, 2002, 245–250.

11. Figovsky, O., Karchevsky, V., and Beilin, D. "Application of Polymeric Coatings for Concrete Structures," *Scientific Israel Technological Advantages* 5, no. 1–2 (2003): 84–92.
12. Weiss, W. "The Behavior of Concrete with Reinforced Plastics Coating," *RILEM*, Paris,1967.
13. Figovsky, O., Karchevsky, V., and Beilin, D. "Load-Carrying Capacity of Polymer Concrete with Polybutadiene Matrix," *J. Scientific Israel Technological Advantages* 4, no. 1 (2002): 21–24.
14. Figovsky, O., Karchevsky, V., and Beilin, D. "Crack-Resistance of Concrete Elements with Polymer Coating," *Proceedings of the Ninth Annual International Conference on Composite Engineering, ICCE/9*, San Diego, CA, 2002, 369–370.
15. Kamaitis, Z. "Structural Design of Polymer Protective Coatings for Reinforced Concrete Structures. Part I: Theoretical Considerations," *Journal of Civil Engineering and Management* 13, No. 1 (2007): 11–17.
16. Kamaitis, Z. "Structural Design of Polymer Protective Coatings for Reinforced Concrete Structures. Part II: Experimental Verification," *Journal of Civil Engineering and Management* 13, no. 1 (2007): 19–26.
17. Surovtsev, I., and Borisov, Yu., "Durability and Deformability of Reinforced Concrete Bending Structures Strengthened by Polymeric Composites," *Journal Scientific Israel Technological Advantages* 14, no. 2 (2012): 12–23.
18. Potapov, Yu., Borisov, Yu., Surovtsev, I., Figovsky, O., and Beilin, D. "Increase of Load-Carrying Capacity of Reinforced Concrete Structures by New Polymer Coating," *Journal Scientific Israel Technological Advantages* 8, no. 1, 2 (Special Issue, 2006): 103–106.

ASTM STANDARDS CITED IN CHAPTER 5

1. **D 56** Standard Test Method for Flash Point by Tag Closed Cup Tester
2. **D 257** Standard Test Methods for DC Resistance or Conductance of Insulating Materials
3. **D 412** Standard Test Methods for Vulcanized Rubber and Thermoplastic Elastomers—Tension
4. **D 471** Standard Test Method for Rubber Property—Effect of Liquids
5. **D 1002** Standard Test Method for Apparent Shear Strength of Single-Lap-Joint Adhesively Bonded Metal Specimens by Tension Loading (Metal-to Metal)
6. **D 1654** Standard Test Method for Evaluation of Painted or Coated Specimens Subjected to Corrosive Environments
7. **D 1729** Standard Practice for Visual Appraisal of Colors and Color Differences of Diffusely Illuminated Opaque Materials
8. **D 2196** Standard Test Methods for Rheological Properties of Non-Newtonian Materials by Rotational (Brookfield type) Viscometer
9. **D 2240** Standard Test Method for Rubber Property—Durometer Hardness
10. **D 2794** Standard Test Method for Resistance of Organic Coatings to the Effects of Rapid Deformation (Impact)
11. **D 4060** Standard Test Method for Abrasion Resistance of Organic Coatings by the Taber Abraser
12. **D 4329** Standard Practice for Fluorescent UV Exposure of Plastics
13. **D 5031** Standard Practice for Enclosed Carbon-Arc Exposure Tests of Paint and Related Coatings
14. **G 151** Standard Practice for Exposing Nonmetallic Materials in Accelerated Test Devices that Use Laboratory Light Sources
15. **G 154** Standard Practice for Operating Fluorescent Ultraviolet (UV) Lamp Apparatus for Exposure of Nonmetallic Materials

6 Epoxy–Rubber Coatings with Nano-Heterogenic Structure

STRUCTURE AND PROPERTIES OF ADVANCED EPOXY–RUBBER COMPOSITION [1–9]

Coatings based on epoxy–amine compositions reveal good chemical resistance, good mechanical properties, and thermostability. Advanced materials for different areas of application can be developed by creation of nano-heterogenic modified epoxy–amine systems with a good specific surface of phase separation. Rubber-modified epoxy coatings exhibit improved fracture toughness as compared to unmodified resins. As was shown [1,2] the properties of rubber-modified epoxy compositions, used as a matrix of anticorrosive coatings, depend not only on the types of epoxy resins, liquid rubbers, and curing agents, but also on the heterogenic structure of these systems. The latter could be characterized by an integral indicator—a specific surface of phase separation.

An integral indicator (n) of structure optimization is determined as follows [3–5]:

$$\ln \tau_{t\sigma} = \ln \tau^{*}_{t\sigma}/\chi^{n} \quad (6.1)$$

where $\tau_{t\sigma}$ is the longevity of the coating, which is determined at the temperature t and at the stress σ, hours; $\tau^{*}_{t\sigma}$ is the longevity of the epoxy–rubber film former of an optimal nanostructure at the same parameters, hours; χ is the phase ratio in epoxy–rubber film former of any and optimal structure; n is the complex indicator, reflecting nonlinear dependence of longevity on phase ratio.

By using conventional mixing technology, the micro-heterogenic systems were obtained. Low molecular weight epoxy resins, such as glycidic ethers of bis-phenol A, bis-phenol F, and other bis-phenols as well as alkylresorcinols, novolac low molecular resins, acrylic low-molecular resins, and so on, in mixtures with aliphatic glycidic ethers having an epoxy equivalent weight (EEW) from 150 to 220 and viscosity from 2,000 to 10,000 cps (25°C) were used. Low molecular weight linear carbon- or hetero-chain liquid rubbers at room temperature, such as butadiene, butadiene-acrylonitril, butadiene-styrene, butadiene-isoprene, chlorisoprene, acrylic and silicone oligomers with terminal reactive carboxyl, hydroxyl, amine and epoxy groups, were used as modifiers. As curing agents, the primary and tertiary amines and aminomethyl phenols were investigated.

In connection with this special interest present phenolic Mannich bases (PMBs) due to two types of reactive groups, such as phenolic hydroxyl and amine, connected with intermolecular hydrogen bonds. Depending on their amino groups, PMBs can be used as hardeners for cold and hot curing.

The advantages of PMBs over traditional amines are as follows:

- Proton donors, such as phenols and alcohols, may not be used, which leads to formation of less sterically volumetric active centers.
- Improvement of adhesive and nonrigid properties of epoxy coatings cured by PMBs due to an increase of network density and a decrease of its inherent flaw.
- Participation of radical products of thermal degradation of PMBs in the epoxy resin curing process at high temperatures.

The influence of the chemical structure of substances in PMB molecules in the curing process of epoxy materials was investigated. Stress–strain properties were determined by traditional physical–mechanical methods. Glass transition temperature was estimated by the thermomechanical method. Chemical resistance of the epoxy based coating cured by PMBs was determined by change of their impact strength after exposition in an aggressive environment within 42 days.

It was established that the reactive ability of PMBs decreases in a row of substitutes—methyl, ethyl, amine, and piperidine groups at an atom of tertiary nitrogen decreases.

With a decrease of the substitution degree of dimethylaminomethyl phenols (DAMPs), the density of the network structure of epoxy resins cured by DAMPs increases. Simultaneous improvement of impact, adhesive and cohesive strength, and chemical resistance of coatings in typical aggressive media takes place (Table 6.1).

TABLE 6.1
Some Properties of Epoxy-bis-phenol-A Resins Cured with DAMPs

Properties	2 DAMP	2, 6 DAMP	2, 4, 6 DAMP
Glass transition temperature (°C)	110	100	90
Impact strength (kJ/m^2)	25	16	11
Tensile strength (MPa)	121	108	86
Coefficient of chemical stability in the media:			
5% NaOH	0.91	0.86	0.82
3% H_2SO_4	0.85	0.83	0.78
5% NaCl	0.90	0.85	0.80
H_2O	0.86	0.84	0.79

Source: Reprinted from O. Blank and E. Gotlib, "Epoxy–Rubber Coatings with Nano-Heterogenic Structure," *J. Scientific Israel Technological Advantages* 10, no. 1 (2008): 32–36. With permission.

So the best stress–strain properties of hot-cured epoxy nanostructured coatings are achieved with use of 2-dimethylaminomethylphenols.

Phenolic Mannich bases in thermooxidizing processes generate active free radicals, namely dialkylaminyl radicals. The latter can participate in opening of epoxy rings by a radical mechanism with formation of carbonyl groups [6]. This increases flexibility of the network structure of epoxy coatings and their impact strength [7].

A better complex of properties of epoxy–rubber materials with PMBs of cold curing is achieved by use of ethylene di aminomethyl phenols (EDAPs). This happens due to the catalytic effect of hydroxyl groups in their molecules.

Coatings based on a mixture of epoxy-diane resins and aliphatic glycidic ethers have improved stress–strain properties. Aliphatic epoxy resins [2] act as an additive the interacts chemically with curing agents and stimulates the formation of block copolymers containing a rigid section of epoxy-diane resins and a flexible one of aliphatic epoxy resin. Oxidation atoms in the network stimulate improvement of impact strength of epoxy materials as compared to nonplasticized ones [3–5]. Simultaneously, network density decreases the same as the speed of curing processes by different types of amines. Structures of nano-heterogenic epoxy–rubber binder obtained by SEM and TEM analyzers are illustrated in Figure 6.1 [4,5].

Epoxy–rubber compositions are homogenic at the stage when components are mixed. During the process of curing of epoxy resins, low molecular weight rubbers form a new elastic disperse phase. The time and degree of phase separation are the same as the volume of elastic disperse phase and the size of its particles (depending on the type of curing agents and chemical activity of liquid rubbers). Low molecular weight rubbers with terminal carboxyl groups are more effective, as demonstrated in Figure 6.2 [9].

By adding new types of surfactants (not more than 1.2% by weight), nano-heterogenic systems with a very good specific surface of phase separation were formed. The data demonstrated the influence of the chemical composition of fluorine-containing surfactants on the properties of epoxy–rubber coatings, and are

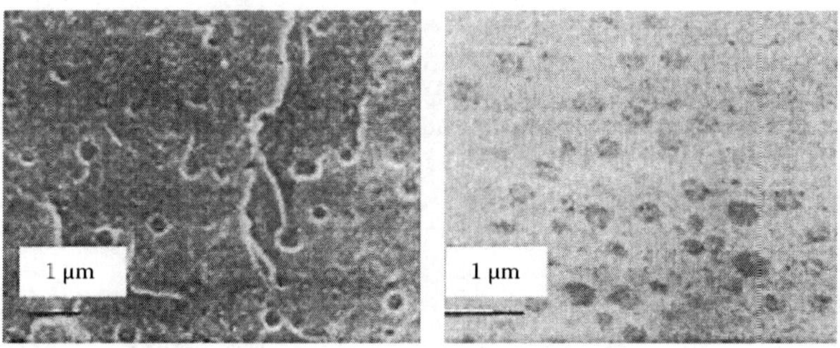

FIGURE 6.1 Microstructures of nano-heterogenic epoxy–rubber binder: (a) epoxy boralkyl resorcin resin modified by nonisocianate polyacetalurethane rubber and hardened by polyethylene polyamines (SEM-analyzer), (b) epoxy-diane resin modified by low molecular fluoroprene and hardened by diamine diphenylhexafluoropropane (TEM-analyzer). (Reprinted from I. Ryblev, N. Blank, and O. Figovsky, "Polymer Compositions Based on Epoxy–Rubber Binders," *Building Materials* no. 6 (1978): 37–39 (in Russian). With permission.)

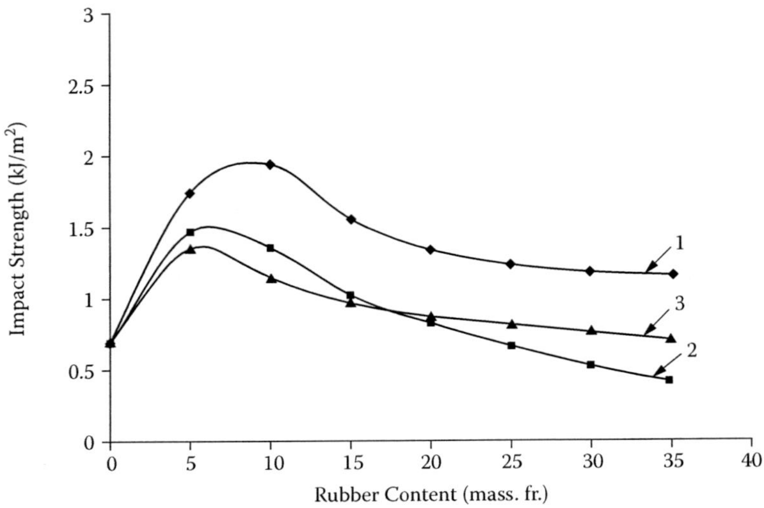

FIGURE 6.2 Dependence of impact strength of epoxy compositions cured by EDAPs from content of liquid rubber with terminal (1) carboxyl, (2) hydroxyl, (3) amine groups. (Reprinted from O. Blank and E. Gotlib, "Epoxy–Rubber Coatings with Nano-Heterogenic Structure," *J. Scientific Israel Technological Advantages* 10, no. 1 (2008): 32–36. With permission.)

presented in Table 6.2. It can be seen that the best strain–strength properties are achieved by use of $CF_3(CF_2)_nCOOH$, where $n = 7–9$ (mixture of homologues). It is connected with its less sterically volumetric molecule.

Smaller improvement of tensile strength takes place when using surfactants with amine and hydroxyl terminal groups. The increase in length of the surfactant chain (the n value) improves the modification effect when the chemical nature of the terminal reactive groups is the same. So surfactants with carboxyl groups (Table 6.2) are more effective. The best relative elongation is achieved by use of surfactants with linear chains and higher n values.

Chemical resistance and thermostability of nano-heterogenic coatings with fluorine surface active additives (FSAA) are approximately 5%–10% higher than for unmodified ones. At the same time, addition of about 3 mass% of metal oxides in epoxy–rubber nanostructured compositions significantly decreases their permeability and increases their chemical resistance in water and water solutions of acids, salts, and alkalis (Figure 6.3) [8].

This is stimulated by selective interaction of described inorganic fillers with aggressive media. Hydrate complexes formed fill macropores and microcracks of coatings. This process results in "repair" of defects and the material is hardened and becomes less penetrable [6,7]. The free volume of nano-heterogenic coatings simultaneously decreases.

Due to increased specific surface of the filler, adhesion on the polymer-additive boundary improves. Penetration of fluid aggressive media in upper layers of coatings usually forms a new mineral phase. This decreases the rate of medium flux and its penetration into the next layers.

TABLE 6.2
Stress–Strain Properties of Nano-Heterogenic Epoxy–Rubber Systems

Fluorine Surface Active Additives (FSAA)	Tensile Strength (MPa)	Relative Elongation (%)	Impact Strength (kJ/m²)
1	2	3	4
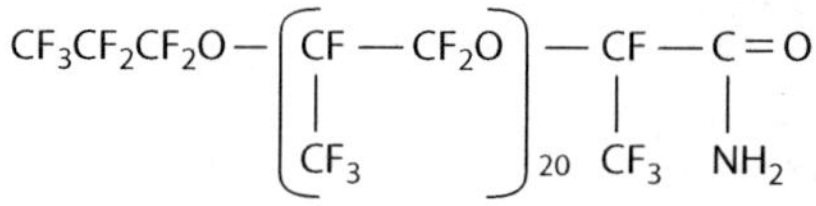	84.8/61.9	21.5/14.2	22.7/14.7
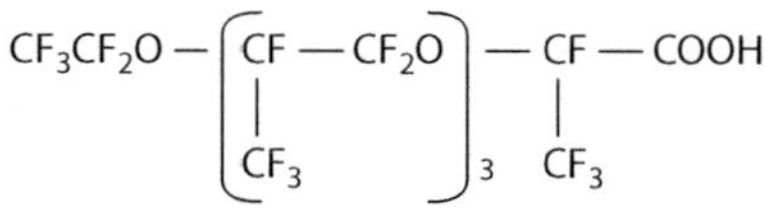	77.4/58.3	19.6/10.1	20.8/12.9
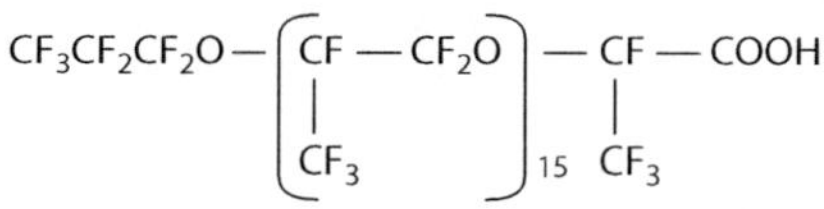	74.2/52.6	22.9/15.2	25.8/17.9
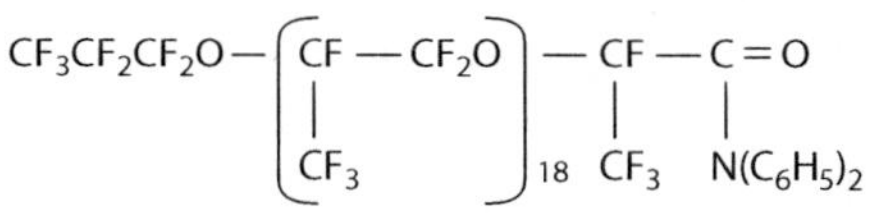	81.8/64.4	22.9/10.8	26.4/15.1
$CF_3(CF_2)_nCOOH$ $n = 7$–9 (mixture of homologues)	93.8/71.7	25.7/11.8	30.4/18.3
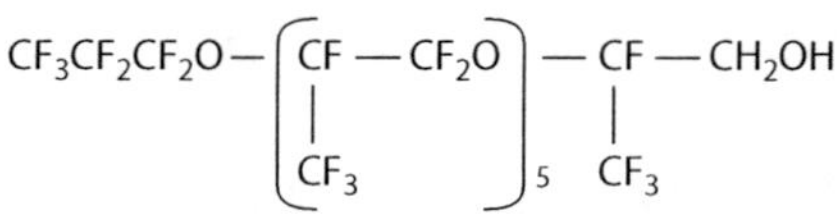	71.9/58.2	22.5/14.9	27.6/18.2
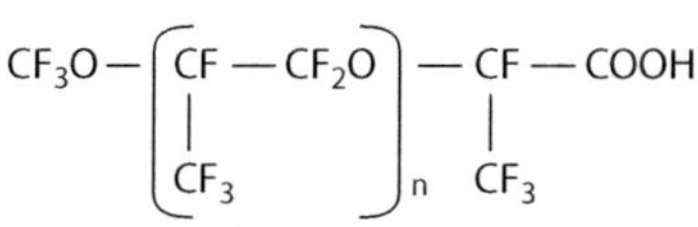			
$n = 10$–12 (mixture of homologues)	74.2/55.4	21.8/15.5	27.2/17.7
$n = 3$–7 (mixture of homologues)	81.0/57.3	20.4/8.7	21.6/10.7
$n = 18$–20 (mixture of homologues)	63.5/50.6	28.9/18.2	29.6/19.4

Source: Reprinted from O. Blank and E. Gotlib, "Epoxy–Rubber Coatings with Nano-Heterogenic Structure," *J. Scientific Israel Technological Advantages* 10, no. 1 (2008): 32–36. With permission.

Note: Numerator data = with FSAA, denominator data = without FSAA

Summarizing the experimental results, we can conclude that formation of nano-heterogenic systems with fluorine-containing surface active additives of optimal chemical composition is an effective method of creating advanced coatings. The mechanical properties and chemical resistance of nano-coatings are significantly higher by use of surfactants with terminal carboxyl groups and less sterically volumetric molecules.

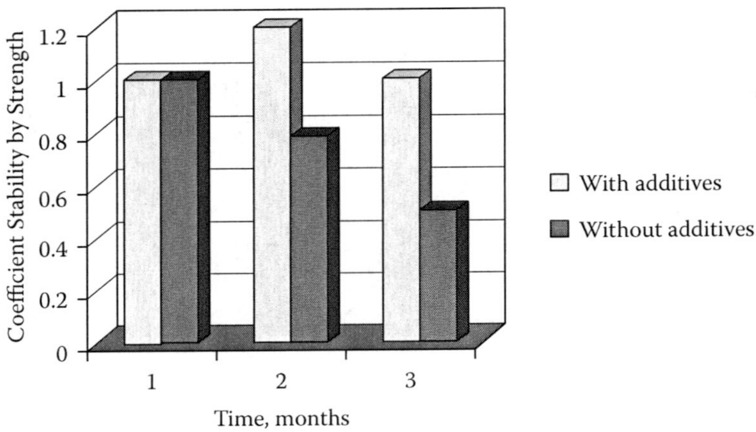

FIGURE 6.3 Chemical resistance of epoxy–rubber coatings in water.

As modifiers, low molecular weight rubbers with carboxyl terminal reactive groups are more effective. As hardeners, we recommend phenolic Mannich bases, such as dimethylaminomethyl phenols for hot curing and ethylene di aminomethyl phenols for cold curing.

Metal oxide nanofillers, which form crystal hydrates with aggressive media, serve as a barrier layer, which prevents deeper penetration of the aggressive medium into coatings and increases their chemical resistance.

REPAIR AND STRENGTHENING OF REINFORCED CONCRETE STRUCTURES BY EPOXY–RUBBER COATINGS [10–25]

Reinforced concrete structures often require strengthening, either due to various damage caused by aggressive environments or by the increasing load that arises during the operation of these structures. The strengthening of reinforced concrete structures by means of externally bonded polymer composite coatings is now considered an effective method for enhancement of durability, load capacity, and crack resistance. Application of such coverings from various polymeric materials is discussed in the following paragraphs.

Experimental research was performed on models of reinforced concrete beams (10 × 20 × 120 cm) with polymeric concrete layers. B 25 class concrete was used; main longitudinal reinforcement of A-III class was used in the form of two 8-mm diameter rods and the 5-mm diameter stirrups of the same class arranged, so that destruction would be along normal sections. Epoxy–rubber compositions were used as covering layer materials. All compositions were made on ED-20® epoxy-based resin. Besides fiber glass–reinforced plastic glued by ED-20 compositions based on liquid rubbers, RubCon was used. The composition of the polymer layers and some of their mechanical characteristics are given in Table 6.3.

The beams were loaded in thirds of the span by consecutively increasing load right up to destruction. The testing scheme is shown in Figure 6.4.

TABLE 6.3
Compositions and Mechanical Properties of the Coating Material

Beam Series	Content of Composites, Particles by Mass	Tensile Strength, MPa	Modulus of Elasticity MPa	Ultimate Tensile Strength, %	Coating Thickness, mm
RCIP	EA-20-100®, DBFT-20®, PEPA-10®	32	2780	1.15	0.15…1.2
RCIM	EA-20-100®, PEPA-10®, DBFT-25®, quartz sand-500	9	11500	0.078	6
RCIS	Glass-reinforced polyether plastic glued by epoxy	22.3	1300	1.7	2
RCIL	EA-20-100®, PEPA-10®, DBFT -30®, crushed sand-91, quartz sand -218, crushed stone-450	7	9000	0.08	Variable
RCIRB	PBN-100®, strengthening group- 75, filler-100, quartz sand- 250, crushed stone-500	14.5	15000	0.86	Variable
RCIRBM	PBN-100®, strengthening group -75, filler-100, quartz sand-500	16	15500	0.91	5…6

Source: Reprinted from O. Blank and E. Gotlib, "Epoxy–Rubber Coatings with Nano-Heterogenic Structure," *J. Scientific Israel Technological Advantages* 10, no. 1 (2008): 32–36. With permission.
Note: DBFT = dibutyl phthalate; PEPA = polyethylenepolyamine; PBN = polybutadienenitrile

The test facility allowed visual monitoring of the development (germination) of cracks in the tension and the side surfaces of the beam. During the test's maximum deflection, deformations of the tension surfaces of the concrete, the average deformation of the concrete over the length of the beam, the local deformations of the concrete in the zone of pure bending, as well as the overall deformation of the zone of pure bending were measured. The measurements enabled determination for each beam of the moment of crack stress, bearing capacity, and deformability of composite beams and the tension stress limit of the concrete.

Influence of Epoxy–Rubber Coating on the Strength of Reinforced Concrete Beams at Bend

The value of the breaking bending moment of the tested composite beams *Mp* is:

$$M_p = M_p^{RC} + M_p^{n} \tag{6.2}$$

where M_p^{RC} is the moment experienced by a reinforced concrete beam without coating and M_p^{n} is the moment experienced by a coating layer only.

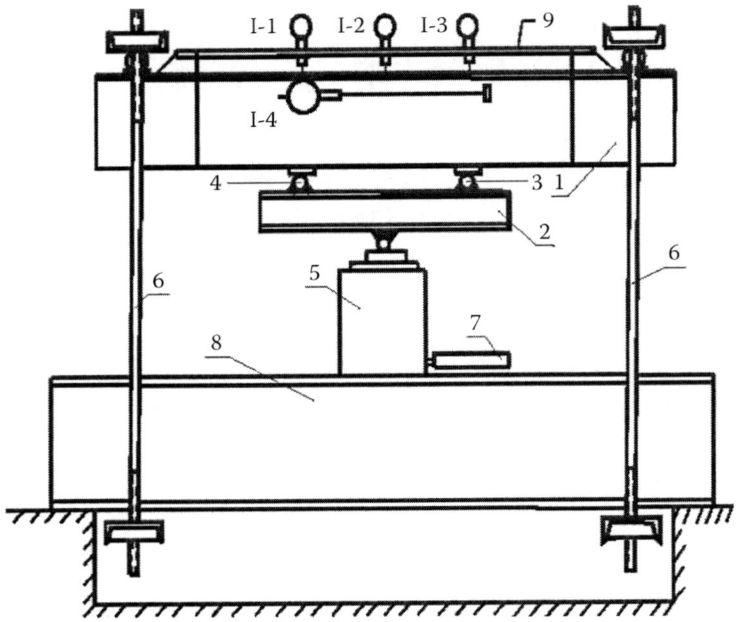

FIGURE 6.4 Testing scheme: (1) beam, (2) distributive traverse, (3, 4) joints, (5) jack, (6) guiding rods, (7) manometer, (8) support beam, (9) holder. (Reprinted from I. Surovtsev, Yu. Potapov, Yu. Borisov, and V. Selyaev, "Strength and Deformability of Reinforced Concrete Bending Constructions Strengthened by Polymeric Composites," *J. Scientific Israel Technological Advantages* 8, nos. 1–2 (2006): 107–116. With permission.)

Let's consider a cross section of a composite beam at bending (Figure 6.5).

$$M_p^{RC} = {}_sA_s\,(h_0 - 0.5x); \tag{6.3}$$

$$M_p^{n} = R_nA_n(h - 0.5x - 0.5\delta_n) \tag{6.4}$$

consequently

$$M_p = R_sA_s(h_0 - 0.5x) + R_nA_n(h - 0.5x - 0.5\delta_n) \tag{6.5}$$

and

$$x = (R_sA_s + R_nA_n) / R_{np} \tag{6.6}$$

where R_n and R_{np} are the tensile strength of the polymer coating and compressive strength of concrete, respectively; R_s is the tensile strength of a reinforcement; A_s and A_n are cross-section areas of reinforcement and polymer coating, respectively; and x is the height of the compressed zone of the composite beam.

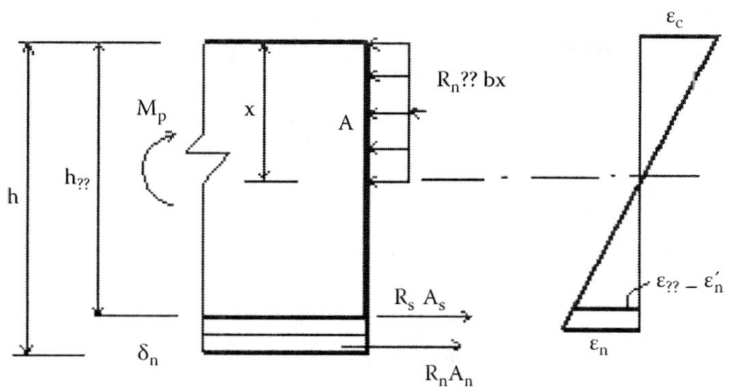

FIGURE 6.5 Stress and deformation diagrams in composite beam section. (Reprinted from I. Surovtsev, Yu. Potapov, Yu. Borisov, and V. Selyaev, "Strength and Deformability of Reinforced Concrete Bending Constructions Strengthened by Polymeric Composites," *J. Scientific Israel Technological Advantages* 8, nos. 1–2 (2006): 107–116. With permission.)

TABLE 6.4
Ultimate Bending Moment Test Results

		Ultimate Bending Moment (kN cm) I			
		Experiment	Calculated Eq. (6.5)		
Thickness of Polymer Coating Layer (mm)	Breaking Load P_p (kN)	M_p^{RCI}	M_p^n	M_p^{exp}	M_p
0	26	415	—	415	430
6	27	415	53	468	446
20	30.5	384	166	550	505
50	43.7	360	370	730	720
90	62.3	336	565	901	1030

Source: Reprinted from I. Surovtsev, Yu. Potapov, Yu. Borisov, and V. Selyaev. "Strength and Deformability of Reinforced Concrete Bending Constructions Strengthened by Polymeric Composites," *J. Scientific Israel Technological Advantages* 8, nos. 1–2 (2006): 107–116. With permission.

Results of the test of the beam series RCIP (Table 6.3) and their comparison with calculations by Equation (6.5) are illustrated in Table 6.4.

Comparison of experimental and calculated values for M_p shows that the difference between them does not exceed 14% when changing the coating thickness from 0 to 90 mm.

The analysis of experimental data shows that the load-carrying capacity of composite beams increases with the thickness growth of the polymer coating layer and its tensile strength. A larger increase of the load-carrying capacity can be achieved by reducing the shrinkage stresses in polymer layer concrete and adding reinforcement to prevent cracks.

Knowing that $A_n = b\delta_n$ where b is the width of the cross-section of the composite beam, the optimal thickness of the polymer coating layer for strength conditions can be obtained from Equations (6.2)– (6.6):

$$\delta_n = (R_{np}h - (R_s A_s / b))/R_{np} + R_n \tag{6.7}$$

Considering that $x = 3.5–5$ cm (Eq. [6.6]), the thickness of the epoxy–rubber coating $\delta_n = 12–16$ cm. This means that the sum of the thickness of the polymer coating layer and the height of the compressed concrete zone is 17–20 cm, which is approximately equal to the total height of the beam (20 cm).

All this testifies to the fact that the maximum effect is obtained with the use of polymer coating for strengthening of reinforced concrete beams and simultaneously for further corrosion protection.

Table 6.5 shows that the load-carrying capacity of the composite beams at bend depends on the strength of the coating material.

Experimental investigations to find the influence of polymer coatings on the strength of concrete and reinforced concrete beams at bend was continued with three series of concrete test specimens (B, B1, and B3) measuring 3 × 6 × 64 cm, and two series of reinforced concrete beam specimens (B0, BP) measuring 10 × 20 × 160 cm. The polymeric coating compositions were EA-20-100®, PEPA-10®, andesite-100®.

Specimens B, B1, and B3 were covered along the lower edge; reinforced concrete specimens B0 and BP were covered on three sides.

The experimental characteristics are presented in Tables 6.6 and 6.7.

It should be noted that the strength of the composite element increases as compared with concrete. For example, for the B1 series, it increased 2–3 times and for B3 it increased more than 3.8 times (Table 6.6). Tests show that the most favorable results were obtained for beams with the polymer concrete layer along three edges.

TABLE 6.5
Test Results: Load-Carrying Capacity of the Composite Beams at Bend

Components of Composition	Breaking Load P_p, (kN)	$(P_p/R_{pn})100$ (%)
ED-20, PEPA, DBTF	87.5	100
ED-20, PEPA, butanol, glass fabric	100	114
ED-20, PEPA, butanol, glass fabric	95	108
ED-20, PEPA, butanol	112.5	128
PBN, Strengthening group	89.5	100
PBN, Strengthening group, glass fabric	102.5	115

Source: Reprinted from I. Surovtsev, Yu. Potapov, Yu. Borisov, and V. Selyaev, "Strength and Deformability of Reinforced Concrete Bending Constructions Strengthened by Polymeric Composites," *J. Scientific Israel Technological Advantages* 8, nos. 1–2 (2006): 107–116. With permission.

TABLE 6.6
Test Results: Influence of Polymer Coatings on the Compressive Strength of Concrete and Reinforced Concrete Beams at Bend

Beam Series	Concrete Grade	Coating Thickness (mm)	Ultimate Compression Stress σ_{nr} (MPa)	Ultimate Bending Moment M_p (Nm)
B	B20	—	2.7	8.33
B 1	B20	1…1.5	6.1	192
B 3	B20	1…1.5	10.5	326
B 0	B12.5	—	—	14000
B P	B12.5	1…1.5	—	16500

Source: Reprinted from I. Surovtsev, Yu. Potapov, Yu. Borisov, and V. Selyaev, "Strength and Deformability of Reinforced Concrete Bending Constructions Strengthened by Polymeric Composites," *J. Scientific Israel Technological Advantages* 8, nos. 1–2 (2006): 107–116. With permission.

TABLE 6.7
Test Results: Influence of Polymer Coatings on the Tension Strength of Reinforced Concrete Beams at Bend

Relative Ultimate Tension Stress	Without Coating	Coating along the Tension and Two Lateral Sides	Coating with Glass Strands	
			Coating along the Tension Side	Coating along the Tension and Two Lateral Sides
(σ_{nr}/R_{np})100%	100	280	304	610

Source: Reprinted from I. Surovtsev, Yu. Potapov, Yu. Borisov, and V. Selyaev, "Strength and Deformability of Reinforced Concrete Bending Constructions Strengthened by Polymeric Composites," *J. Scientific Israel Technological Advantages* 8, nos. 1–2 (2006): 107–116. With permission.

Note: Reinforced concrete beam samples measuring 140 × 20 × 10 cm from B 15 class concrete, reinforced with A-III class longitudinal reinforcement of 20-mm diameter and A-II class stirrups of 6 mm.

Influence of Epoxy–Rubber Coating on Deformability of Reinforced Concrete Beams at Bend

The influence of a polymer concrete layer on the deformability of reinforced concrete beams was studied by testing beams measuring 140 × 20 × 10 cm from B-15 class concrete, reinforced with 20-mm diameter A-III class longitudinal reinforcement and 6-mm i diameter A-II class stirrups at three bend points.

Five species of composite beams were tested:

1. Type RCR had no coating.
2. Type RC2E1 had epoxy composition coating along the tension side of the beam.

3. Type RC2E3 had epoxy composition coating along the tension and two lateral sides of the beam.
4. Type RC2ESI had epoxy polymer composition strengthened with glass strands along the tension side of the beam.
5. Type RC2ES3 had epoxy polymer composition strengthened with glass strands along the tension and two lateral sides of the beam.

The deflection in the middle of the span was taken as the deformability criterion and fixed at loads of 40 kN. Three stages of deformation were distinguished at the loading beam with coating:

- The beginning of loading until the moment the cracks appear in the concrete
- From the moment the cracks appear until covering fracture
- Beam destruction

On the boundary of the *first and the second* deformation stages, the polymer concrete layer has not been destroyed and performs its main function: it isolates the concrete and reinforcement from the aggressive environment. Crack formation and its development are observed on the boundary of the *second and the third* deformation stages right up to break.

The quantitative deformability characteristics of beams at bend is given in Table 6.8. As seen in Table 6.9, composite beams have smaller deflections f as compared to the reinforced concrete beam without a strengthened coating.

The research shows that polymer concrete coatings reduce the deformability of reinforced concrete beams, especially before cracks appear in concrete (i.e., on the boundary of the first and second stages of deformability). The greatest effect of deformability decrease is observed when coatings are put along three edges. Reduction of deformability of the beams with a small thickness polymer coating layer may be explained by the redistribution of stresses between the coating and

TABLE 6.8
Deflection of the Reinforced Concrete Beam Samples at Bend

Type of Beam	Deflection during the Instant of Crack Formation f_{cf} (mm)	Deflection at Load $P = 40$ kN f (mm)	f_{RC} (without coating)
RCR (without coating)	0.62	1.2	1
RC2E1	0.60	1.07	0.89
RC2E3	1.06	1.06	0.88
RC2ESI	0.88	1.13	0.94
RC2ES3	1.34	0.9	0.75

Source: Reprinted from I. Surovtsev, Yu. Potapov, Yu. Borisov, and V. Selyaev, "Strength and Deformability of Reinforced Concrete Bending Constructions Strengthened by Polymeric Composites," *J. Scientific Israel Technological Advantages* 8, nos. 1–2 (2006): 107–116. With permission.

TABLE 6.9
Deflection of Beams at Bend

Type of Beam	Service Stress σ_0 (Pa)	Specific Creep at 400 Days f_{400}/σ_0	Creep Coefficient f_{400}/f_0
B	1.7	3.18	30
	1.5	2.80	32
	1.0	3.00	40
	0.7	3.40	48
B1	3.8	0.39	3.6
	3.0	0.43	3.0
	2.2	0.47	2.2
	1.5	0.54	1.5
	1.1	0.57	1.1
	0.75	0.54	0.8
B3	7.0	0.21	2.9
	6.2	0.19	3.3
	5.1	0.17	3.3
	4.0	0.17	2.7

Source: Reprinted from I. Surovtsev, Yu. Potapov, Yu. Borisov, and V. Selyaev, "Strength and Deformability of Reinforced Concrete Bending Constructions Strengthened by Polymeric Composites," *J. Scientific Israel Technological Advantages* 8, nos. 1–2 (2006): 107–116. With permission.

Note: f_{400} = *deflection after 400 days at service stress* σ_0, f_0 = *initial deflection at the same stress.*

concrete during loading, with the result being a "smoothing" of peak strains on the tension side of the concrete and reinforcement takes place.

During formation of cracks in concrete, deflection of the beam samples with a polymer coating layer (the second stage of deformation) increased sharply and approximates the deflection of the control samples (more elastic than concrete). Deformability increases with the growth of the polymer concrete layer thickness (Figure 6.6).

The influence of polymer coatings on the deformability at long-term action of bending loads (creep) was studied on the same specimens: beams series B (plain concrete), B1 (with coating along the lower edge), and B3 (with coating along the lateral and lower edges).

As is seen from the creep curves (Figures 6.7–6.11), deformability of composite beams is much lower than that of concrete beams. For example (Figure 6.10), at stress of 1.9 MPa under short-term bending loading, deflections of B series specimens (plane concrete beams) and those of composite beams of B1 and B3 series, respectively, are equal to 0.19, 0.13, and 0.1 mm; under long-term loading (more 200 days) these deflections are equal to 0.6, 0.43, and 0.6.

From Table 6.9 it follows that specific creep at 400 days and the creep coefficient for composite beam samples (B1 and B3) are one order of magnitude less than for plain concrete.

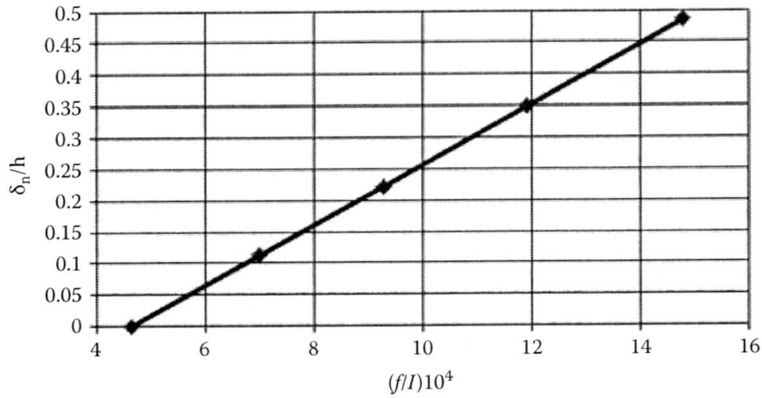

FIGURE 6.6 Dependence of polymer coating thickness (δ_n) in tension zone on deflection of the bend composite beams (*l* is the span of the beam, *h* is the height of the cross section. (Reprinted from I. Surovtsev, Yu. Potapov, Yu. Borisov, and V. Selyaev, "Strength and Deformability of Reinforced Concrete Bending Constructions Strengthened by Polymeric Composites," *J. Scientific Israel Technological Advantages* 8, nos. 1–2 (2006): 107–116. With permission.)

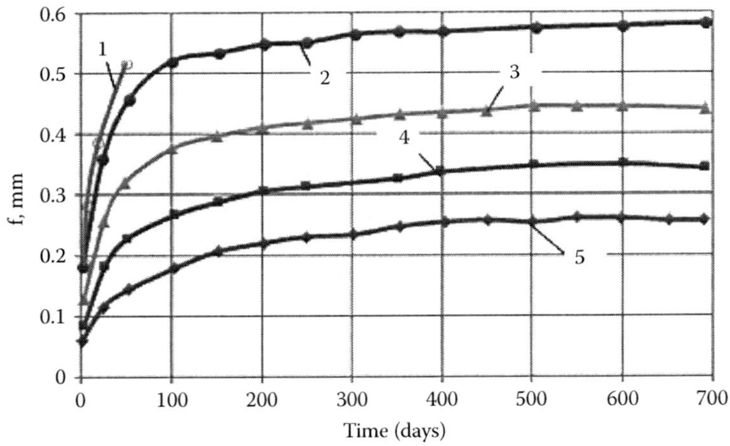

FIGURE 6.7 Creep curves of plain concrete beam samples (B0) at bend: (1) σ_0 = 2.2 MPa, (2) σ_0 = 1.7 MPa, (3) σ_0 = 1.5 MPa, (4) σ_0 = 1.0 MPa, (5) σ_0 = 0.70 MPa. (Reprinted from I. Surovtsev, Yu. Potapov, Yu. Borisov, and V. Selyaev, "Strength and Deformability of Reinforced Concrete Bending Constructions Strengthened by Polymeric Composites," *J. Scientific Israel Technological Advantages* 8, nos. 1–2 (2006): 107–116. With permission.)

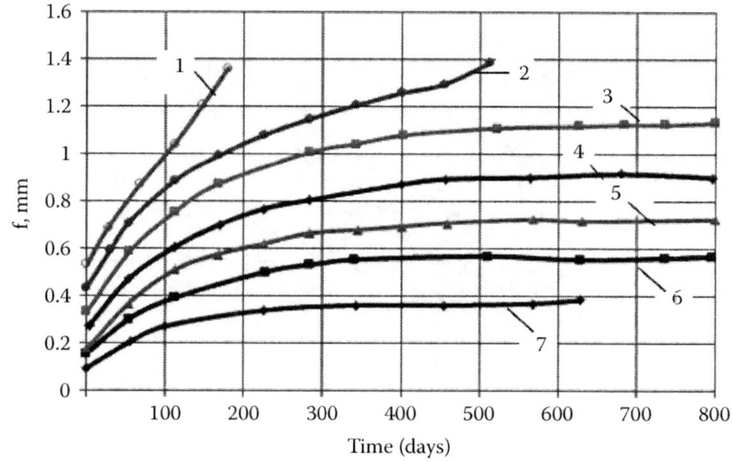

FIGURE 6.8 Creep curves of composite concrete beam samples (B1) with coating along the lower side at bend: (1) σ0 = 4.6 MPa, (2) σ0 = 3.8 MPa, (3) σ0 = 3.0 MPa, (4) σ0 = 2.2 MPa, (5) σ0 = 1.5 MPa, (6) σ0 = 1.1 MPa, (7) σ0 = 0.75 MPa. (Reprinted from I. Surovtsev, Yu. Potapov, Yu. Borisov, and V. Selyaev, "Strength and Deformability of Reinforced Concrete Bending Constructions Strengthened by Polymeric Composites," *J. Scientific Israel Technological Advantages* 8, nos. 1–2 (2006): 107–116. With permission.)

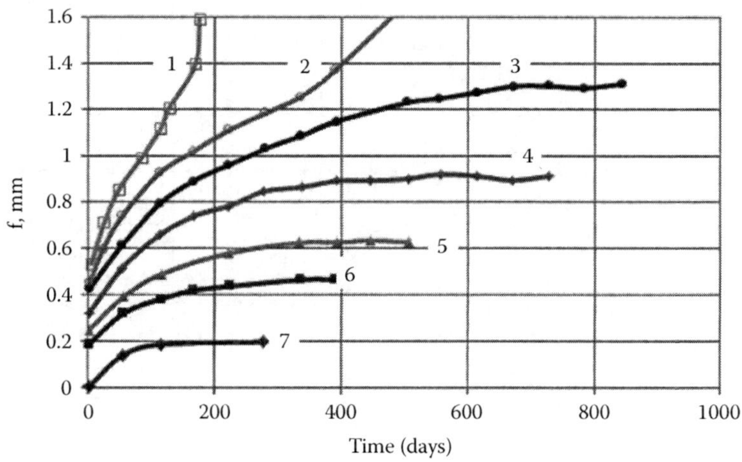

FIGURE 6.9 Creep curves of composite concrete beam samples (B2) with coating along three sides at bend: (1) σ0 = 8.0 MPa, (2) σ0 = 7.0 MPa, (3) σ0 = 6.2 MPa, (4) σ0 = 5.1 MPa, (5) σ0 = 4.0 MPa, (6) σ0 = 3.0 MPa, (7) σ0 = 1.5 MPa. (Reprinted from I. Surovtsev, Yu. Potapov, Yu. Borisov, and V. Selyaev, "Strength and Deformability of Reinforced Concrete Bending Constructions Strengthened by Polymeric Composites," *J. Scientific Israel Technological Advantages* 8, nos. 1–2 (2006): 107–116. With permission.)

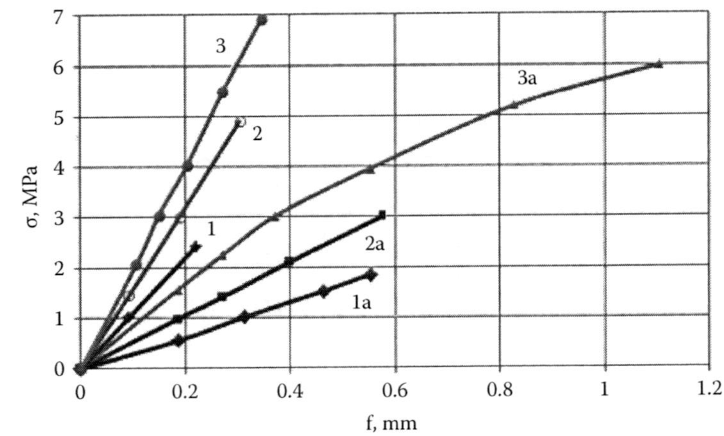

FIGURE 6.10 Dependence of deflections on stress at short (full line) and long-term (dotted line) loading for beam samples at bend: (1, 1a) plain concrete, (2, 2a) with coating along the lower side, (3, 3a) with coating along three sides. (Reprinted from I. Surovtsev, Yu. Potapov, Yu. Borisov, and V. Selyaev, "Strength and Deformability of Reinforced Concrete Bending Constructions Strengthened by Polymeric Composites," *J. Scientific Israel Technological Advantages* 8, nos. 1–2 (2006): 107–116. With permission.)

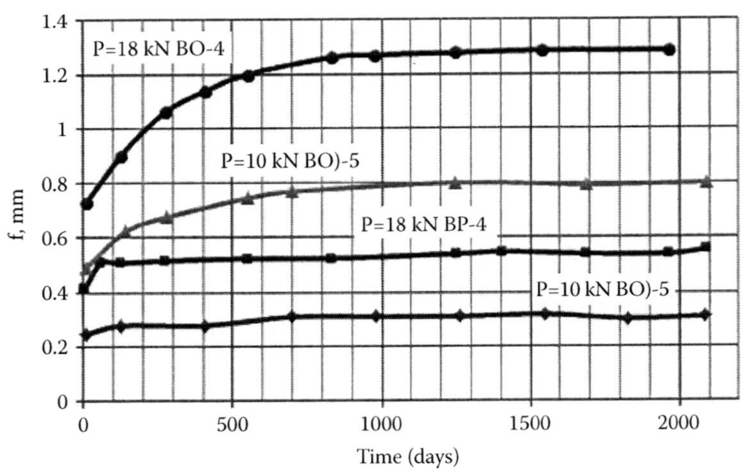

FIGURE 6.11 Creep curves of reinforced plane concrete and composite beams. (Reprinted from I. Surovtsev, Yu. Potapov, Yu. Borisov, and V. Selyaev, "Strength and Deformability of Reinforced Concrete Bending Constructions Strengthened by Polymeric Composites," *J. Scientific Israel Technological Advantages* 8, nos. 1–2 (2006): 107–116. With permission.)

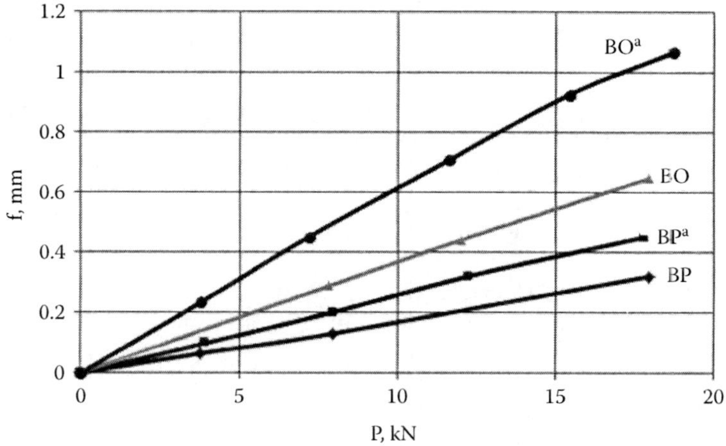

FIGURE 6.12 Dependence of deflections on loading for plain reinforced concrete (B0) and composite (BP) beams at short-term loading (BO, BP) and at long-term loading (B0^{a}, BP^{a}). (Reprinted from I. Surovtsev, Yu. Potapov, Yu. Borisov, and V. Selyaev, "Strength and Deformability of Reinforced Concrete Bending Constructions Strengthened by Polymeric Composites," *J. Scientific Israel Technological Advantages* 8, nos. 1–2 (2006): 107–116. With permission.)

Tests of reinforced elements for a long-term load were carried out in specially designed plants that allowed two beam samples to be tested simultaneously: one control beam and one with a coating.

Loading was realized by equal concentrated forces at thirds of the span, and beam deflections were measured in the middle of the span and on the boundaries of the pure bending zone.

Using the two beam samples, the control reinforced concrete beam (series B0) and beams with coating (series BP) were tested (Figure 6.12). It should be noted that B0-4 and BP-4 samples were tested at concentrated force P = 18 kN, which corresponds to the load at which cracks appear in short-term tests for beams without a coating. In so doing, in testing of the B0-4 sample, two hairline cracks appeared in the zone of pure bending; in sample B0-4 with a coating, no cracks were found.

The curves of beam creep are given in Figure 6.10 and Figure 6.11.

Formation of cracks and their development in time are fixed at the very beginning of the test of the B0-4 samples. The width of the crack openings were 0.3–0.35 mm/year. Cracks were also formed under the coating, but the coating itself retained its integrity up to the tests' completion.

SUMMARY

- Formation of nano-heterogenic systems with fluorine-containing surface active additives of optimal chemical composition is an effective method of obtaining advanced coatings.

- The mechanical properties and chemical resistance of nano-coatings are significantly higher when using surfactants with terminal carboxyl groups.
- Tests show that the epoxy–rubber coating strength of reinforced concrete beams reduces deformability of the reinforced concrete beams in bending, especially before crack formation in concrete; the most favorable results are obtained for beams with a polymer concrete layer on three sides.

REFERENCES

1. Gotlib E., Litvina T., and Sokolowa, J. "The Investigation of Epoxy Polymers Modification by Rubbers with Carboxyl Groups," *Visokomol. Soed.* B 17, no. 10 (1975): 793–795 (in Russian).
2. Gotlib E., Sokolowa J., and Kisiljeva, R. "Zur modifizierung von epoxiharsen mit reactionsfagien elastomeren," *Plaste und Kautsch.* 27, no. 4 (1980): 191–194.
3. Moshinsky, L., and Figovsky, O. "Chemical Resistance of Epoxy Polymers Depending on a Structure of Amino-Phenol Hardeners," *Scientific Israel-Technological Advantages* 1, no. 1 (1999): 28–34.
4. Ryblev, I., Blank, N., and Figovsky, O. "Polymer Compositions Based on Epoxy–Rubber Binders," *Building Materials* no. 6 (1978): 37–39 (in Russian).
5. Blank, N., and Figovsky, O. "Epoxy–Rubber Coating with Nano-Heterogenic Structure," *Kraski i Laki,* no. 10 (2009): 14–16 (in Russian).
6. Kirpichnikov, P. A., Averko-Antonovich, Yu. A., and Averko-Antonovich, L. A. "Chemistry and Technology of Synthetic Rubber," Khimia, Moscow 1987 (in Russian).
7. Sokolowa, J., and Gotlib, E. *Composition Materials Based on Modified Polymers.* Moscow: Uniarprint, 2000, 1998 (in Russian).
8. Figovsky, O., and Romm, F. "Improvement of Anti-Corrosion Protection Properties of Polymer Materials," *Anti-corrosion Methods & Materials* 45, no. 3 (1998): 312–330.
9. Figovsky, O., and Dubrovsckaya-Vinokurova, G. "Shock Resistance of Monolithic Multilayer Flooring," *Scientific Israel Technological Advantages* 5, no. 2 (2003): 165–169.
10. Figovsky, O., and Beilin, D. "Building Materials Based on Advanced Polymer Matrix: Review," *Scientific Israel Technological Advantages* 10, no. 2 (2008): 1–119.
11. Potapov, Yu., Solomatov, V., and Selyaev, V. *Polymer Coatings for Reinforced Concrete Structures*, Moscow: Stroiyizdat, 1973 (in Russian).
12. Figovsky, O., Beilin, D., Blank, N., Potapov, Yu., and Chernyshev, V. "Development of Polymer Concrete with Polybutadiene Matrix," *J. Cement & Concrete Composites* 18, no. 6 (1996): 437–444.
13. Figovsky, O., and Shapovalov, L. "Nanostructured Hybrid Nonisocyanate Polyurethane Coatings," *J. Paint and Coatings Industry* no. 6 (2005): 36–44.
14. Surovtsev, I., Potapov, Yu., Borisov, Yu., and Selyaev, V. "Strength and Deformability of Reinforced Concrete Bending Constructions Strengthened by Polymeric Composites," *J. Scientific Israel Technological Advantages* 8, no. 1–2 (2006): 107–116.
15. Potapov, Yu., Borisov, Yu., Surovtsev, I., Figovsky, O., and Beilin, D. "Increase of Load Capacity of Reinforced Concrete Structures by New Polymer Coating," *J. Scientific Israel Technological Advantages* 8, no. 1–2 (2006): 103–107.
16. Weiss, V. *The Behavior of Concrete with Reinforced Plastics Coating*. Paris: RILEM, 1967.
17. Figovsky, O., Karchevsky, V., and Beilin, D. "Protective Crack-Resistant Waterborne Coatings Based on Vulcanized Chlorine-Sulpho-Polyethylene," *Scientific Israel Technological Advantages* 3, no. 1–2 (2001): 121–125.

18. Beilin, D., and Figovsky, O. "Stress–Strain State of Polymer Chemical-Resistant Coatings Curing and under Operating Conditions," *J. Composite Structures* 31 (1995): 253–256.
19. Figovsky, O., Karchevsky, V., and Beilin, D. "Crack-Resistant and Anticorrosive Coatings Based on Vulcanized Water Dispersion of Chlorine-Sulpho-Polyethylene," *J. Anticorrosion Methods and Materials* 50 (2003): 108–114.
20. Kamaitis, Z. "Structural Design of Polymer Protective Coatings for Reinforced Concrete Structures. Part I: Theoretical Considerations," *Journal of Civil Engineering and Management* 13, no. 1 (2007): 11–17.
21. Depuy, G. "Polymer Concrete Overlays for the Repair and Protection of Concrete," *ICPIC 2001. The Tenth International Congress in Polymeric Concrete,* Honolulu, HI, 2001.
22. Figovsky, O., Karchevsky, V., and Beilin, D. "High-Performance Waterborne Protective Coatings Based on Curable Water Dispersion of Chlorine-Sulphonated Polyethylene," *Proceedings of the International Paint & Auxiliary Products Industry Congress & Exhibition BOYA/PAINT 2008,* Istanbul, Turkey, 2008, 115–125.
23. Beilin, D., and Figovsky, O. "Strengthening of Concrete Structures by the Advanced Crack-Resistant Coatings Based on Water Dispersion CSPE," *Book of Abstracts of First International Conference RAR-2006,* Voronezh, Russia, November 9–10, 2006, 55–57.
24. Figovsky, O., Karchevsky, V., and Beilin, D. "Stress–Strain State of Elastomeric Crack-Resistant Coatings for Concrete Substrates," *Book of Abstracts.* June 9–13, 2002, Riga, Latvia, 29.
25. Potapov, Yu., Borisov, Yu., Surovtsev, I., Figovsky, O., and Beilin, D. "Increasing Repair and Crack Resistance of Reinforced Concrete Structures by Polymer Composites," *Proceedings of the Second International Conference on Concrete Repair, Concrete Solution*, St. Malo, France, 2006, 111–114.
26. Blank, O., and Gotlib, E. "Epoxy–Rubber Coatings with Nano-Heterogenic Structure," *J. Scientific Israel Technological Advantages* 10, no. 1 (2008): 32–36.

7 Nanostructured Binder for Acid-Resistant Building Materials

Acid-resistant building materials based on liquid glass find wide application in construction as silicate polymer concretes, filler pasties, putties, and so on. The application of soluble sodium silicates (liquid glasses) is associated with their use as the binding components for the manufacture of heat- and chemical-resistant materials. Liquid glasses have high cohesive strength, resistance to corrosion, do not release flammable volatile components, are low in cost, and are ecologically friendly material.

Significant increase in strength, and resistance to heat and fire of a silicate matrix was achieved by introducing tetrafurfuryl esters of orthosilicic acid (tetrafurfuryloxisilane [TFS]) in the composition. The effect is achieved by strengthening of contacts between silica gel globules and a modified alkaline component due to "inoculations" of the furan radical. Introduction of the TFS additive in the binding medium leads to formation of nanoparticles of SiO_2, which act as nucleation centers of crystallization, and furfural alcohol, which fills the silicate matrix and forms a cross-linked polymer. The TFS additive increases mechanical strength and chemical resistance of the binder and is widely used for preparation of acid-resistant concretes and void fillers [1,2].

COMPOSITION OF THE NOVEL NANOSTRUCTURED BINDER

This new nanostructured binder was obtained by laminar mixing of the liquid glass containing alkali metal cations as Na, TFS,* and a water-soluble silicate, containing an organic alkaline cation (WSS) such as 1,8-diazabicycloundecene-7 (DBUS, Structure 7.1) or 1,5-diazabicyclononene-5 (DBNS, Structure 7.2).

The composition of these materials is shown in Table 7.1.

Water-soluble silicate, containing an organic alkaline cation, can be obtained by interaction of quaternary organic ammonium base salts and amorphous silica. The relationship of SiO_2 to the organic base is 1.65:1 for DBUS and 2.0:1 for DBNS. It should be noted that the soluble organic alkali silicates, for example, tetrabutylammonium silicate (TBAS) is also used as a component of self-extinction binder.

* Nanostructuring additive tetrafurfuryloxisilane(TFS) $Si(O_2C_5H_5)_4$ is the reaction product of ethyl esters of orthosilicic acid and furfuryl alcohol. Tetraethoxysilane $(C_2H_5O)_4Si$ - ethyl ester of orthosilicic acid contents SiO_2 in terms of silica 60 wt%.

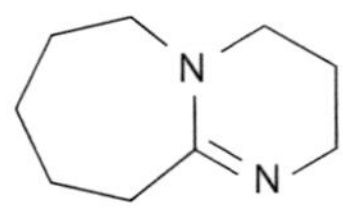

STRUCTURE 7.1 DBUS ($C_9H_{16}N_2$).

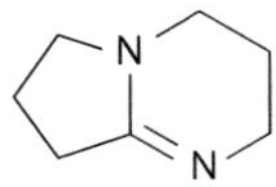

STRUCTURE 7.2 DBNS ($C_7H_{12}N_2$).

TABLE 7.1
Composition of the Novel Nanostructured Binder

Component	% (by mass)
Sodium silicate (Liquid glass), containing alkali metal cautions as Na	80–95
Tetrafurfuryloxisilane	2–7
Water-soluble silicate, containing an organic alkaline caution	2–4

Source: Reprinted from O. Figovsky, Yu. Borisov, and D. Beilin, "Nanostructured Binder for Acid-Resisting Building Materials," *Journal Scientific Israel Technological Advantages* 14, no. 1 (2012): 7–12. With permission.

After mixing of all components, the binder must be used within 2–3 hours. Hardener was introduced with fine ground mineral filler.

For experimental verification of the proposed binder properties, various compositions of building materials were prepared (Table 7.2):

- Acid-resistant silicate concrete
- Acid-resistant patching material

Acid-resistant silicate concrete and patching materials were prepared by mixing the components in the proportions shown in Tables 7.3 and 7.4. The test results of acid-resistant silicate polymer samples are shown in Table 7.5 (also see Figure 7.1–7.3).

One can see that the compressive strength of silicate polymer concretes modified by additive WSS increases by 30%, and water permeability is reduced by more than 15 times in comparison with concrete without the additive.

Table 7.6 and Figures 7.4 and 7.5 illustrate the test results of acid-resistant void fillers based on the new binder.

TABLE 7.2
Experimental Composition of the Binder

Composition Number	Component	% (by mass)
1	Liquid glass	95
	TFS	3
	WSS (DBUS)	2
2	Liquid glass	88
	TFS	8
	WSS (DBUS)	4
3	Liquid glass	92
	TFS	5
	WSS (DNNS)	3
4	Liquid glass	83
	TFS	12
	WSS (DNNS)	5

Source: Reprinted from O. Figovsky, Yu. Borisov, and D. Beilin, "Nanostructured Binder for Acid-Resisting Building Materials," *Journal Scientific Israel Technological Advantages* 14, no. 1 (2012): 7–12. With permission.

TABLE 7.3
Composition of Acid-Resistant Silicate Polymer Concretes

		% (by mass)			
		Concrete 1	Concrete 2	Concrete 3	Concrete 4
Components of Concrete Mixes				Without WSS	
Binder: composition	No. 3	13.4	—	13.4	—
	No. 1	—	13.2	—	13.2
Sodium silicate (Ssp = 2700 cm^2/g)		1.8	1.7	1.8	1.7
Powder of diabase (Ssp = 3000 cm^2/g)		18.0	18.0	18.0	18.0
Quartz sand (Msize = 2.4)		28.0	28.0	28.0	28.0
Granite macadam		38.8	39.1	38.8	39.1

Source: Reprinted from O. Figovsky, Yu. Borisov, and D. Beilin, "Nanostructured Binder for Acid-Resisting Building Materials," *Journal Scientific Israel Technological Advantages* 14, no. 1 (2012): 7–12. With permission.

TABLE 7.4
Compositions of Acid-Resistant Void Filler Mixtures

Components of Patching Materials		% (by mass) Void Filler Mixture 1	Void Filler Mixture 2	Void Filler Mixture 3 (Without WSS)	Void Filler Mixture 4 (Without WSS)
Binder: composition	No. 2	41.0	—	41.0	—
	No. 4	—	34.0	—	34.0
Sodium silicate (Ssp = 2700 cm^2/g)		6.0	6.2	6.0	6.2
Powder of diabase (Ssp = 3000 cm^2/g)		53.0	—	53.0	—
Powder of andesite (Ssp = 2600 cm^2/g)		—	59.8	—	59.8

Source: Reprinted from O. Figovsky, Yu. Borisov, and D. Beilin, "Nanostructured Binder for Acid-Resisting Building Materials," *Journal Scientific Israel Technological Advantages* 14, no. 1 (2012): 7–12. With permission.

TABLE 7.5
Test Results of Acid-Resistant Silicate Polymer Concrete Samples

Properties		% (by mass) Concrete 1	Concrete 2	Concrete 3 (Without WSS)	Concrete 4 (Without WSS)
1		**2**	**3**	**4**	**5**
Ultimate strength at compression, MPa (Figure 7.1)		49.2	48.7	37.1	35.4
Water permeability, 10^{-7} cm^2/g (Figure 7.2)	After 7 days	0.42	0.48	8.77	8.96
	After 30 days	0.14	0.16	6.10	6.44
1		**2**	**3**	**4**	**5**
Coefficient of acid resistance after 360 days exposure to these environments:					
5% HCl		1.03	1.02	0.97	0.93
20% HCl		1.12	1.07	0.96	0.91
5% H_2SO_4		1.06	1.05	0.93	0.94
20% H_2SO_4 (Figure 7.3)		1.17	1.18	0.92	0.90

Source: Reprinted from O. Figovsky, Yu. Borisov, and D. Beilin, "Nanostructured Binder for Acid-Resisting Building Materials," *Journal Scientific Israel Technological Advantages* 14, no. 1 (2012): 7–12. With permission.

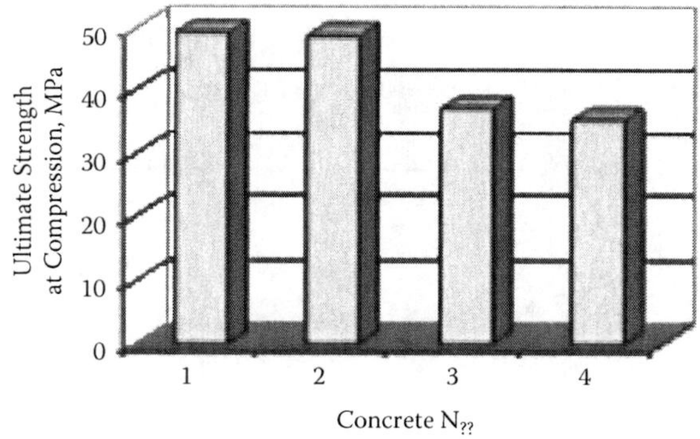

FIGURE 7.1 Ultimate strength at compression, MPa. (Reprinted from O. Figovsky, Yu. Borisov, and D. Beilin, "Nanostructured Binder for Acid-Resisting Building Materials," *Journal Scientific Israel Technological Advantages* 14, no. 1 (2012): 7–12. With permission.)

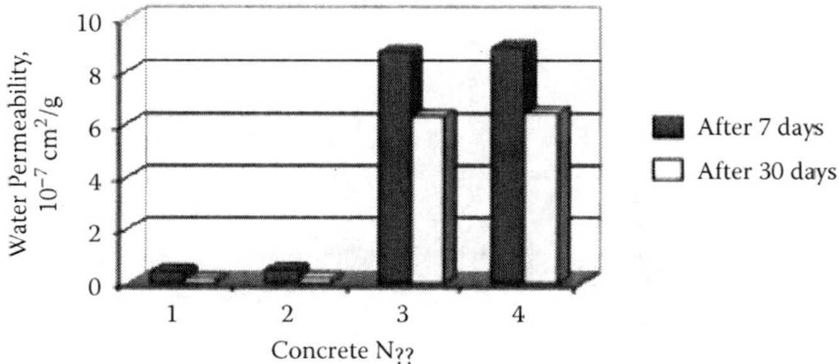

FIGURE 7.2 Water permeability, 10^{-7} cm^2/g. (Reprinted from O. Figovsky, Yu. Borisov, and D. Beilin, "Nanostructured Binder for Acid-Resisting Building Materials," *Journal Scientific Israel Technological Advantages* 14, no. 1 (2012): 7–12. With permission.)

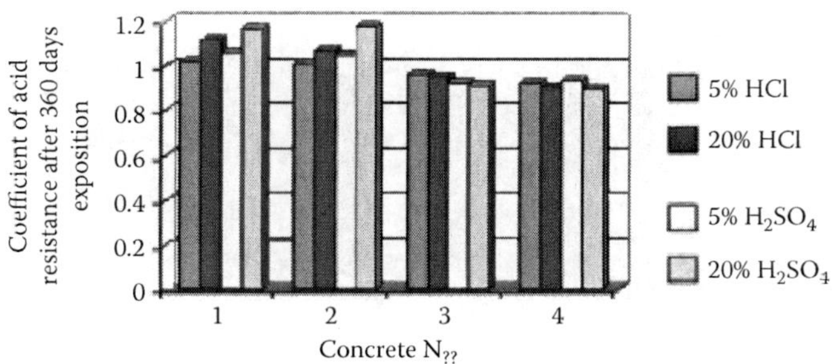

FIGURE 7.3 Coefficient of acid resistance after 360 days exposition into the aggressive environments. (Reprinted from O. Figovsky, Yu. Borisov, and D. Beilin, "Nanostructured Binder for Acid-Resisting Building Materials," *Journal Scientific Israel Technological Advantages* 14, no. 1 (2012): 7–12. With permission.)

TABLE 7.6
Test Results of Acid-Resistant Void Filler Samples

Properties	% (by mass)			
	Void Filler 1	Void Filler 2	Void Filler 3	Void Filler 4
			Without WSS	
Adhesive strength at shear, MPa:				
• Ceramic–steel	1.54	1.48	0.92	0.95
• Ceramic–ceramic	1.90	1.94	1.18	1.22
(Figure 7.4)				
Adhesive strength at tension, MPa				
• Ceramic–steel	0.98	1.04	0.47	0.51
• Ceramic–ceramic	1.27	1.30	0.55	0.54
(Figure 7.5)				

Source: Reprinted from O. Figovsky, Yu. Borisov, and D. Beilin, "Nanostructured Binder for Acid-Resisting Building Materials," *Journal Scientific Israel Technological Advantages* 14, no. 1 (2012): 7–12. With permission.

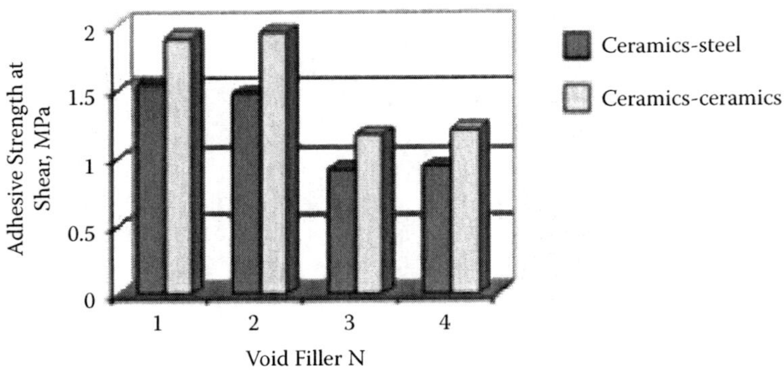

FIGURE 7.4 Adhesive strength at shear. (Reprinted from O. Figovsky, Yu. Borisov, and D. Beilin, "Nanostructured Binder for Acid-Resisting Building Materials," *Journal Scientific Israel Technological Advantages* 14, no. 1 (2012): 7–12. With permission.)

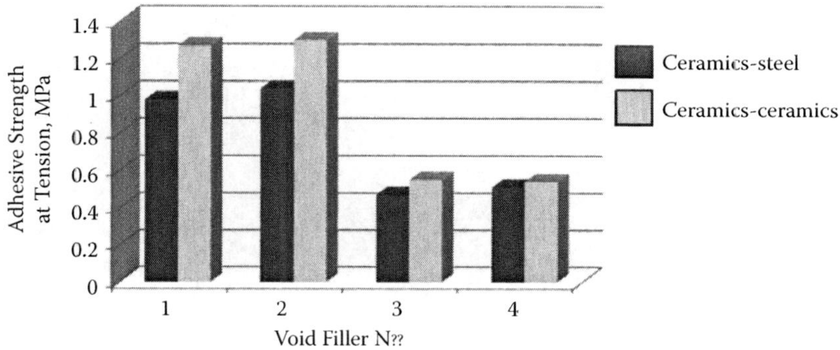

FIGURE 7.5 Adhesive strength at tension. (Reprinted from O. Figovsky, Yu. Borisov, and D. Beilin, "Nanostructured Binder for Acid-Resisting Building Materials," *Journal Scientific Israel Technological Advantages* 14, no. 1 (2012): 7–12. With permission.)

SUMMARY

- Test results illustrate that adhesive strength at shear of acid-resistant void filler modified by additive WSS increases by 55% for the ceramic–steel combination and by 50% for the ceramic–ceramic combination in comparison with the void filler without the additive.
- Adhesive strength at tension increases by 20% for the ceramic–steel combination and more than 2 times for the ceramic–ceramic combination in comparison with the void filler without the additive.

REFERENCES

1. Figovsky, O., and Beilin, D., "Improvement of Strength and Chemical Resistance of Silicate Polymer Concrete," *International Journal of Concrete Structures and Material*, 3, no. 2 (2009): 97–101.
2. Beilin, D., Borisov, Yu., Figovsky, O., and Surovtsev, I., "Nanostructured Binder for Composite Building Materials," Patent Russian Federation, RU 2408552 C1, 2011.
3. Figovsky, O., and Beilin, D., "Optimal Composition, Strength, and Chemical Resistance of Silicate Polymer Concrete," *Proceeding of the 13th International Congress on Polymers in Concrete*, February 10–12, 2010, Funchal-Madeira, Portugal, 142–151.
4. Figovsky, O., Borisov, Yu., and Beilin, D., "Nanostructured Binder for Acid-Resisting Building Materials," *Journal Scientific Israel Technological Advantages* 14, no. 1 (2012): 7–12.

8 Waterborne Fire-Protective and Heat-Stability Coating Compositions

Application of fire-resistant coatings is one of the basic methods of protecting wood and other flammable materials from the influence of a fire and increased temperature. It is desirable to impart both short-term protection, to allow evacuation from a building, and long-term protection, to protect the surfaces from heat and fire damage and to prevent combustion. In recognition of the foregoing, various compositions have been developed that can add fire-retardant (-resistant) performance to substrates.

Known multicomponent fire-protective coating compositions such as [1,2,3] have a number of serious limitations. Many of these coatings require the application of toxic organic solvents and fire retardants, such as antimony oxide or lead octoate, that release suffocating gases while burning and pollute the environment. Often such compositions require the application of multiple layers consisting of different compositions, of which at least one is toxic. Many waterborne fire-protective coatings (especially transparent ones) that are manufactured on a base of water-soluble inorganic materials become brittle and covered with whitish spots after long-term exposure to atmospheric conditions.

We formulated waterborne, nontoxic, environmentally friendly coatings that impart flame- and fire-resistant properties while maintaining good weatherproofing characteristics including UV radiation resistance. The coatings may be used for fire protection or heat stability of a number of substrates, both flammable and inflammable as lignocellulosic and polymeric materials, such as solid lumber, parallel strand lumber, timber strand, plywood, particle board, medium-density fiberboard, hardboard, oriented strand board and straw board, gypsum boards, fiberglass, metals, ceramics, and so on. The composition provides a transparent or colored decorative coating that is capable of providing heat and fire protection for interior as well as exterior applications.

COATING COMPOSITIONS

The developed coating is a waterborne fire-protective composition consisting of a combination of intumescent organic and inorganic particles, an inorganic water glass, a water dispersion of chlorine-sulfonated polyethylene (CSPE) and pigments (or silicon dioxide).

The intumescent organic and inorganic particles in the composition act to retard burning by expanding upon exposure to heat and/or flame and endothermically releasing a stream–gas mixture at the temperature of inflammation. The steam–gas release temperature depends upon the composition of the expandable organic and inorganic particles. For example, aluminum hydroxide releases water vapor at 180°C –200°C while sodium oxalate released CO_2 at a temperature of about 450°C. Thus, in preferred embodiment, mixtures of organic and inorganic particles are used to impart temperature protection. Suitable intumescent or heat-expandable particles include aluminum hydroxide and organic salts of alkali metals, such as sodium oxalate, having a particle size no larger than 90 μk.

The inorganic water glass after hardening has a melting temperature (550°C) just above the temperature at which the steam–gas mixture is released from the intumescent particles. Thus, after the intumescent organic and inorganic particles have expanded, a further increase in temperature will cause an endothermic release of water from water glass and formation of a foamable vitrified barrier coating. This vitrified barrier operates to separate the coated substrate from the oxygen supply and cut off the oxygen needed to support combustion. Furthermore, the inorganic water glass may react with the intumescent particles, in particular with aluminum hydroxide particles, resulting in the formation of alumosilicates, which provide a synergetic effect and increase the fire-protective properties of the coating. Any suitable inorganic water glasses can be used to produce the coatings, such as sodium or potassium silicate.

CSPE is an elastomer; due to its elasticity, it imparts flexibility to the coating, preventing internal stress creation as evidenced by cracking and crazing of the applied coating. We have found that although elastomers decrease fire resistance, the particular chlorinated polymer CSPE in water dispersion contains about 40%–50% dry material mass and actually *increases* the fire resistance of the coating composition. During exposure to heat, CSPE forms a protective carbonated layer that separates the substrate from the oxygen supply. This layer, together with a vitrified layer, increases the fire protection barrier effect. Additionally, CSPE imparts both moisture and weather resistance to the coating, particularly resistance to atmospheric carbon dioxide.

The pigments and silicon dioxide act to thicken the coating composition to allow easy coating on both vertical and horizontal surfaces. The pigments impart coloration to the coating, and silicone dioxide provides the desired degree of transparency. For production of a transparent coating, an amorphous silicon dioxide, such as fumed silica, is added to the coating. For colored coatings, an appropriate inorganic pigment, or mixture of pigments, such as TiO_2, Cr_2O_3, Fe_2O_3, ZnO, and the like, is added to the coating to achieve the desired color.

For production of a heat-stable coating on a metal substrate, *inorganic fibers and glass microbubbles* are added to the coating composition described above in place of the pigment or the silica. Inorganic fibers are mullite from groups Fiberfax® groups (average fiber diameter 1.5–2.5 μkm) or analogous. Glass bubbles are 10–70 μm in diameter and white in color. The aforementioned ingredients provide a heat barrier due to the porous structure of the coating composition and effectively retard heat propagation. These ingredients decrease the average coating density and reduce temperature conductivity of a coating as well.

TABLE 8.1
Composition of the Coatings (%)

Kind of Coating			
M EXT	**M ASB**	**S EXT**	**S ASB**
Heat-Insulating Coating		**Fire-Protective Coating**	
Outdoor Application	**Indoor Application**	**Outdoor Application**	**Indoor Application**
Sodium silicate liquid	Sodium silicate liquid	Sodium silicate liquid	Sodium silicate liquid
Aluminum hydroxide	Aluminum hydroxide	Aluminum hydroxide	Aluminum hydroxide
Sodium oxalate	Sodium oxalate	Sodium oxalate	Sodium oxalate
Borax	Borax	Titanium (IV) oxide	Titanium (IV) oxide
Nonorganic fibers	Nonorganic fibers	Latex CSPE	(or other inorganic pigment)
Glass bubbles	Glass bubbles		
Perlite/vermiculite	Perlite/vermiculite		
Latex CSPE			

The composition of the two kinds of coatings for indoor and outdoor application, heat insulating (MEXT, M ASB) and fire protective (S EXT S ASB), are shown in Table 8.1.

The coating components are blended in appropriate tanks in a manner known in the industry. The ingredients are mixed in a suitable mixer such as a ball or glass bead mixer. All of the ingredients except the water dispersion of CSPE are first mixed together and then CSPE is mixed in.

The resultant fire-protective and heat-insulating coatings can be applied to the substrate by brush, roller, and by airless or pneumatic spraying. Preferably 2 to 4 coats are applied to the surface, with each coat being dried 1–2 hours before spreading of the next coat. The total thickness of the applied transparent coating is between 0.2 and 0.4 mm; for colored or opaque coatings, the number of coating layers may be increased to 4–6 coats, with concomitant increases in the total coating thickness. For heat-stability coating composition, containing inorganic fibers and glass bubbles, the total thickness of applied coats is 1.0–2.0 mm. The coated surface is then cured for 1 to 2 days outdoors at a temperature of at least 20°C and a relative humidity of less than 65%.

PHYSICAL–MECHANICAL PROPERTIES OF THE NEW FIRE-PROTECTIVE AND HEAT-INSULATING COATING COMPOSITIONS

Technical characteristics of the fire-protective and heat-insulating coatings are shown in Table 8.2.

Fire-Protective Coating for Wood Substrates

Wood samples measuring 150 × 30 mm were coated with 4 layers of fire-protective coating compositions S ASB for indoor application by brush with layer-by-layer

TABLE 8.2
Technical Characteristics of Coatings

	Kind of Coating			
	M EXT	M ASB	S EXT	S ASB
	Heat-Insulating Coating		Fire-Protective Coating	
Main Properties	Outdoor Application	Indoor Application	Outdoor Application	Indoor Application
Substrate	Metal, concrete, and other noncombustible materials		Wood, plastic, and other combustible substrate transferring them in a nonflammable class of materials	
Density	1.3 g/cm^2	1.5 g/cm^2	1.8–1.9 g/cm^2	
pH	10.5–11.5			
Viscosity (ASTM D2196)	70000–90000 cPs	18000–25000 cPs	70000–90000 cPs	1400–1500 cPs
Adhesion	To metal: 3.3–3.5 MPa To EFPS ≥ 3 MPa			
Drying time	24 hours at room temperature or 6 hours at 60°C (layer by layer)			
Moister resistance	1.7%–3.5% for inflammable substrates			
Class of fire-resistance (Life Safety Code, NFPA 101)			Class A (ASTM E84-00a)a Flame spread index = 10 Smoke development index = 15	
Time of reaching 500°C for metal substrate	60–100 min (ASTM E119)[b]			
Time of reaching 100°°C for EFPS substrate	No less than 30 minutes			
Toxicity	Both compositions are nontoxic.			
Decay/Rot resistance	Both compositions are decay- and rot-resistant materials.			

[a] Test was carried out by Underwriters Laboratory and Testing Service of Hardwood, Plywood and Veneer Association (Reston, Virginia 20190, USA)

[b] Test was carried out by Laboratory of Ormat Ltd., Israel.

drying of 1–2 hours. The thickness of the resulting coating was 0.45 mm. Coated and cured substrates were tested in accordance with ASTM E119.

Decease in the mass decrease of fire-protective coatings at exposure over the open flame of a gas burner for 2 minutes at a temperature of 200°C and a flame height of 15 cm was measured. The charring time of wood samples tested over a flame at a temperature of 950°C–1050°C was measured. These conditions simulate fire conditions that cause inflammation of a wood substrate.

Unprotected and protected wood samples after exposure in fire are shown in Figure 8.1.

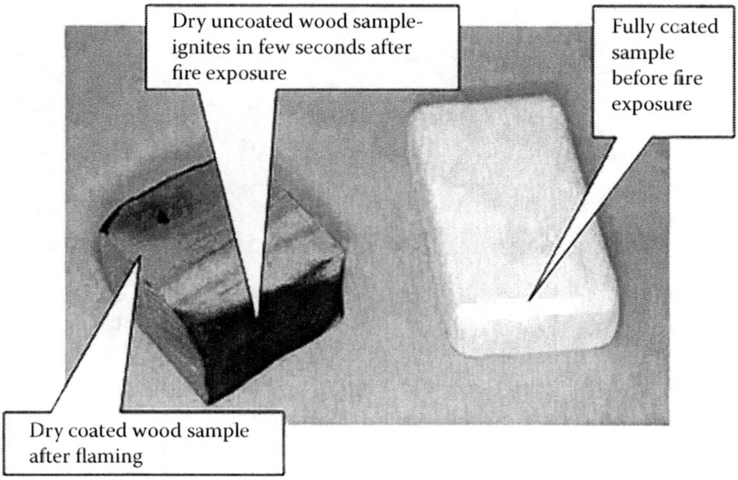

FIGURE 8.1 Coated wood sample before (right) and after fire exposure test. (Reprinted from O. Figovsky, V. Karchevsky, and D. Beilin, "Advanced Hybrid Fire-Protective Coatings," *J. Scientific Israel Technological Advantages* 4, nos. 3–4 (2002): 94–97. With permission.)

Fire tests in a 25-foot tunnel furnace were carried out according to standard ASTM E84 by the Underwriters Laboratory and the Hardwood Plywood and Veneer Association. Results are given for the flame spread index (FSI) and smoke developed index (SDI). The values obtained from burning the test materials represent a comparison with that of ¼ inch inorganic reinforced cement board expressed as zero and red oak flooring expressed as 100.

Test samples were fir boards 244 cm in length using three sections measuring 51 × 1.3-cm wide by 244-cm long for a total specimen length of 732 cm, coated with fire-protective composition S ASB for indoor application. Maximal temperature in the tunnel furnace was approximately 370°C.

The flame spread distance was observed and recorded every 15 seconds or every 2 feet of progression. The peak distance was noted at the time of occurrence. The flame spread distance was plotted over time. The total area under the flame spread distance–time curve was determined. The flame spread was then calculated as a function of the area under the curve relative to the standard red oak curve area. The value for flame spread classification was compared with that of inorganic reinforced cement board and select-grade red oak flooring.

The smoke developed during the test is determined by the reduction in output of a photoelectric cell. The value of smoke developed is derived by calculating the net area under the curve for the test material and comparing this area with net area under the curve for uncovered red oak flooring.

Results of the tests from the Underwriters Laboratory report are shown in Figures 8.2 and 8.3.

Results of tests have shown that the offered composition conforms to Class A according to Safety Code NFPA 101 [8] (FDI = 10; SDI = 15).

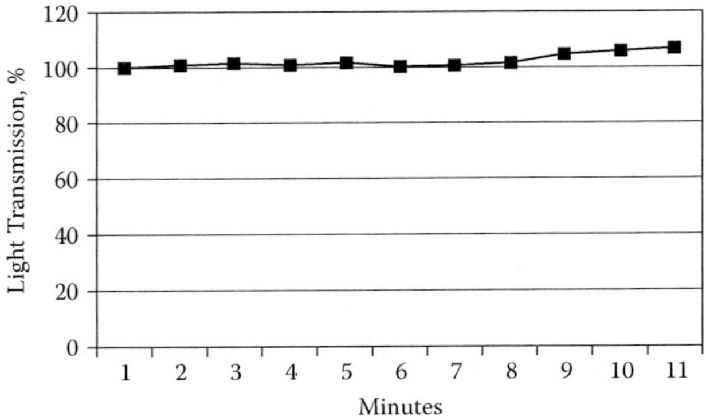

FIGURE 8.2 Smoke development during the fire test. (Reprinted from O. Figovsky, V. Karchevsky, and D. Beilin, "Advanced Hybrid Fire-Protective Coatings," *J. Scientific Israel Technological Advantages* 4, nos. 3–4 (2002): 94–97. With permission.)

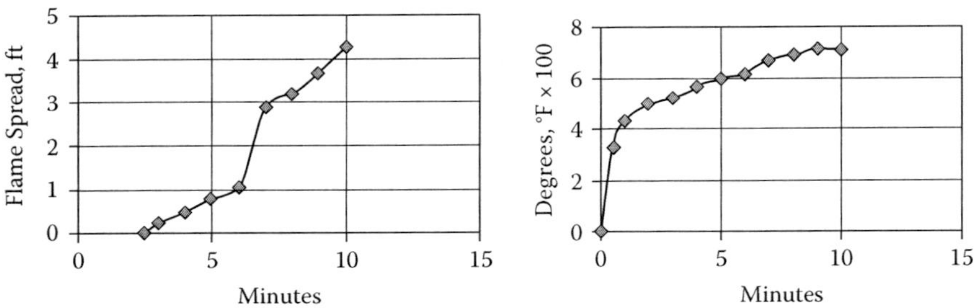

FIGURE 8.3 Fire spreading during the fire test. (Reprinted from O. Figovsky, V. Karchevsky, and D. Beilin, "Advanced Hybrid Fire-Protective Coatings," *J. Scientific Israel Technological Advantages* 4, nos. 3–4 (2002): 94–97. With permission.)

Fire-Protective Coating for Extruded Foam Polystyrene

Extruded foam polystyrene (EFPS) is used in construction as heat insulation material, leave-in-place forms, and so on. However, insufficient temperature resistance prevents its widespread use as a structural building material.

EFPS samples measuring 135 × 70 × 20 mm were coated with the S ASB composition for indoor application. All sides were coated with two layers of total thickness of 2–2.5 mm with level-by-level drying in the furnace at a temperature of 60°C for 4 hours. Before the tests began, the samples were incubated for 24 hours at room temperature and humidity.

Tests of the EFPS samples were performed in the laboratory furnace " Carbolite" supplied by the digital thermometer. Five samples with a covering and 3 control samples without a covering were tested. Heating of the samples took place with a smooth growth of temperature in the furnace: on the a <150°C with a speed ~ 10°C/min, on a site > 150°C with a speed ~ 5°C/min. Registration of the heating temperature of

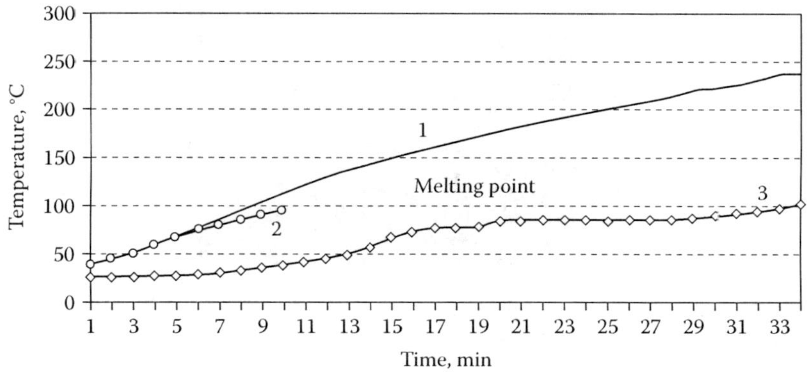

FIGURE 8.4 Dependence of the temperature measured inside the EFPS samples on heating time: (1) sample coated by composition S ASB for indoor application, (2) sample without coating, (3) temperature in the furnace.

FIGURE 8.5 The EFPS samples: (a) sample without coating, (b) sample with coating, (c) and (d) samples after fire testing.

the sample was done by a thermocouple inside the sample body after every minute of heating [7].

Figure 8.4 shows a typical dependence of the temperature measured inside the sample on heating time and the temperature plot in the furnace.

The melting point of the control sample without a coating (100°C) is reached after 10 minutes, whereas the same melting temperature of the sample with S ASB coating is reached more than 30 minutes after the beginning of heating.

Emission of combustion gases were not observed during the test. The gases are bounded chemically by the coating composition and through encapsulation create a rigid heat-resistant barrier shell around the sample.

A general view of the EFPS samples without the covering and with a covering after fire testing is shown in Figures 8.5 and 8.6.

Heat-Insulating Coating for Steel Substrate

Steel plates measuring 600 × 600 × 1.5 mm were coated on one side with coating composition M ASB, indoor application, which is designed for metal substrates in thicknesses of 1.4–1.7 mm, and tested in a furnace according to ASTM E119. Temperature was registered on the uncoated side of the samples. Temperature curves are illustrated in Figure 8.7.

FIGURE 8.6 Cross-section of the EFPS sample after the fire test. One can see a molten sample within the rigid shell casing.

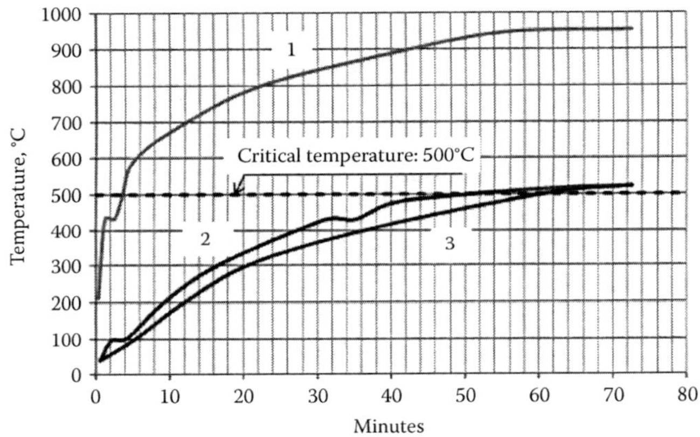

FIGURE 8.7 Results of fire tests of coated steel samples: (1) temperature in furnace, (2) and (3) samples.

It is easy to see the critical temperature: 500°C was reached after 60–100 minutes of heating; for control samples (without coating) this temperature was attained after 5–7 minutes.

Moisture Resistance of Fire-Protective and Heat-Insulating Coatings for Outdoor Application

Coated and cured wood and steel substrates were tested for moisture resistance at 100% relative humidity for 1, 7, and 28 days in accordance with ASTM D2247 to determine the amount of moisture absorbed by the coating. Results of the test are illustrated in Table 8.3.

Adhesion Resistance of Fire-Protective and Heat-Insulating Coatings

Wood and steel samples coated with two kinds of fire-protective and heat-insulating coatings were subjected to testing in accordance with ASTM D1002 to determine

TABLE 8.3
Compositions and Properties of Fire-Protective Coatings

Coating Composition	Substrate	Mass Decrease (%)	Time (min)	Moisture Resistance, (% absorption) 1 day	7 days	28 days
Fire-protective S EXT	Wood	2.0	More than 15[a]	0.9	2.1	3.0
Heat-insulating M EXT	Steel	1.8	100[b]	1.1	2.2	3.8

Source: Reprinted from O. Figovsky, V. Karchevsky, and D. Beilin, "Advanced Hybrid Fire-Protective Coatings," *J. Scientific Israel Technological Advantages* 4, nos. 3–4 (2002): 94–97. With permission.

[a] Time of exposure in naked flame until to inflammation of wood substrate

[b] Time of exposure in naked flame until to 500°C of steel substrate.

the adhesion characteristics of the coating to the substrate. Adhesion of the coating to the wood substrate is 3.1±0.5 MPa; adhesion of the coating to the steel substrate is 3.3±0.5 MPa.

SUMMARY

- New fire-protective and heat-insulating waterborne coating compositions are nontoxic, noncombustible, and nonflammable.
- The production and use of the coating compositions are environmentally friendly and do not cause ecological problems.
- These coating compositions are characterized by high weather resistance; additionally, the transparent coating not only provides fire protection, but also preserves the visible wood structure to create decorative wood surfaces.
- These compositions can be manufactured at the plant in accordance with the varnish-paint standard technology and be supplied to the user as a finished product ready for application to the substrates to be protected.

REFERENCES

1. Scholz, G. and Pitig, W-D. Expandable, flame-retardant coating compositions, US Patent 5,749,948.
2. Vajs, L., and Pettit, D. Fire retardant, US Patent 4,663,226, Issued: May 5, 1987.
3. Ellis, H. Fire barrier coating composition containing magnesium oxychlorides and high alumina calcium aluminate cements or magnesium oxysulphate, US Patent 4,572,862.
4. Figovsky, O., Karchevsky, V., and Beilin, D. "Advanced Hybrid Fire-Protective Coatings," *J. Scientific Israel Technological Advantages* 4, nos. 3–4 (2002): 94–97.
5. Figovsky, O., Karchevsky, V., Beilin, D., and Aksenov, O. "Waterborne Fire Protective Coating Composition," *Abhandlungen Der WIGB*, Band 3, Berlin (2003): 74–78.

6. Figovsky, O., Karchevsky, V., Beilin, D., and Aksenov, O. "Advanced Waterborne Fire Protective and Heat Retarded Coating Compositions," *Conference Papers, Organic-Inorganic Hybrids II: Science, Technology, Applications*, University of Surrey, Guilforg, UK, 28–29 May 2002, Paper no. 16.
7. Romm, F., Karchevsky V., and Figovsky L. "Optimal Fire Tests of Protective Coatings," *Materials Performance* no. 1 (2002): 44–47.
8. Safety Code NFPA 101.

ASTM STANDARDS CITED IN CHAPTER 8

1. **E 84-00a** Standard Test Method for Surface Burning Characteristics of Building Materials
2. **E 119** Standard Test Methods for Fire Tests of Building Construction and Materials
3. **D 2196** Standard Test Methods for Rheological Properties of Non-Newtonian Materials by Rotational (Brookfield type)Viscometer

Index

D

E

M

N

T

U

V